Solving Nonlinear Equations with Iterative Methods

Fundamentals *of* Algorithms

Editor-in-Chief: Nicholas J. Higham, University of Manchester

The SIAM series on Fundamentals of Algorithms is a collection of short user-oriented books on state-of-the-art numerical methods. Written by experts, the books provide readers with sufficient knowledge to choose an appropriate method for an application and to understand the method's strengths and limitations. The books cover a range of topics drawn from numerical analysis and scientific computing. The intended audiences are researchers and practitioners using the methods and upper level undergraduates in mathematics, engineering, and computational science.

Books in this series not only provide the mathematical background for a method or class of methods used in solving a specific problem but also explain how the method can be developed into an algorithm and translated into software. The books describe the range of applicability of a method and give guidance on troubleshooting solvers and interpreting results. The theory is presented at a level accessible to the practitioner. MATLAB® software is the preferred language for codes presented since it can be used across a wide variety of platforms and is an excellent environment for prototyping, testing, and problem solving.

The series is intended to provide guides to numerical algorithms that are readily accessible, contain practical advice not easily found elsewhere, and include understandable codes that implement the algorithms.

Series Volumes

C. T. Kelley

North Carolina State University
Raleigh, North Carolina

Solving Nonlinear Equations with Iterative Methods

Solvers and Examples in Julia

siam | Society for Industrial and
Applied Mathematics

Publications Director	Kivmars H. Bowling
Executive Editor	Elizabeth Greenspan
Managing Editor	Kelly Thomas
Production Editor	Louis R. Primus
Copy Editor	Paul Bernard
Production Manager	Donna Witzleben
Production Coordinator	Cally A. Shrader
Compositor	Cheryl Hufnagle
Graphic Designer	Doug Smock

Library of Congress Cataloging-in-Publication Data
Names: Kelley, C. T. author.
Title: Solving nonlinear equations with iterative methods : solvers and
    examples in Julia / C.T. Kelley, North Carolina State University, Raleigh,
    North Carolina.
Description: Philadelphia : Society for Industrial and Applied Mathematics,
    [2023] | Preceded by: Solving nonlinear equations with Newton's method /
    C.T. Kelley. c2003. | Includes bibliographical references and index.
Identifiers: LCCN 2022029161 (print) | LCCN 2022029160 (ebook) | ISBN
    9781611977264 (paperback) | ISBN 9781611977271 (ebook)
    Subjects: LCSH: Iterative methods (Mathematics) | Nonlinear theories. | Julia
    (Computer program language)
Classification: LCC QA297.8 .K457 2023 (ebook) | LCC QA297.8 (print) | DDC
    518/.26 23/eng20220--dc02
LC record available at *https://lccn.loc.gov/2022029161*

To Betty Thomas Kelley

Contents

Preface

This book on solvers for nonlinear equations is a user-oriented guide to algorithms and implementation. It is a sequel to [111], which used MATLAB for the solvers and examples. This book uses Julia [17] and adds new material on pseudo-transient continuation, mixed precision solvers, and Anderson acceleration.

Roughly a third of each of the first three chapters is taken from [111]. The new material includes new and more detailed examples, pseudo-transient continuation, new codes for the solvers and examples, and all the source codes in GitHub repositories. Chapter 4 on Anderson acceleration and the case studies in Chapter 5 are completely new.

Each of the first four chapters has a component created as a print book in LATEX and a final section created as an IJulia (Jupyter) notebook. I combined the two formats in slightly different ways in the print book and notebook versions, so the correspondence between the two is not exact. The fifth chapter on case studies was written as a print book and then mapped into the notebook.

The purpose of the book is to show, via algorithms in pseudocode and Julia with several examples, how one can choose an appropriate iterative method for a given problem and write an efficient solver or apply one written by others.

This book is intended to complement my older book [107], which focuses on in-depth treatment of convergence theory, but does not discuss the details of solving particular problems, implementation in any particular language, or evaluating a solver for a given problem.

The computational examples in this book were done with Julia 1.7.2 on various Apple computers. The Julia codes for the solvers are in a Julia package [115]. I provide a suite of Jupyter notebooks [114] to enable the reader to run all the examples in the book and play with them.

I have found that Julia is an excellent environment for a project like this, which unifies a print book, a package of solvers, and an interactive notebook. I have used all the solvers in my own research, and many of the examples come from that research.

I have written the codes with a goal of readability by a Julia novice who knows some numerical analysis. In particular, I have made the algorithmic parameters very easy to find. I have also put the methods for managing the iteration such as the line search, error reporting, and the logic for updating Jacobians and preconditioners in a directory that is easy for the interested reader to find and for the less interested reader to ignore. I have sacrificed a good deal of abstraction, some generality, and a bit of performance for clarity.

The solvers were designed for production work on desktop and laptop computers to solve small- to medium-scale problems having at most a few tens of thousands of unknowns. Very large scale problems on leadership-class computers are best done with software (Trilinos [86], PETSc [12], SunDials [91]) designed for that purpose.

We assume that the reader has a good understanding of elementary numerical analysis at the level of [11] and of numerical linear algebra at the level of [53, 99, 195]. A student who has had, or is taking, a course from one of these books is well prepared. We will also use some deeper results from numerical linear algebra and will refer the reader to [90] for those.

The examples are closely coupled to the text, and the reader will get the most out of this book if she/he has an elementary knowledge of Julia. If the reader has good skills in numerical work in another high-level language, then learning Julia at the level one needs to work through this book is not difficult and can be done while reading the book.

The reader should know (or learn!) Julia and its ecosystem well enough to use the package manager to install packages, use modules, start a notebook, and do basic tasks in numerical methods (LU, SVD, QR) in Julia. You should also know how to use GitHub to clone repositories and put them where they need to go. A reader with that minimal knowledge of GitHub and Julia should be able to install and understand the codes, play with the algorithms, and wreak havoc.

Parts of this book are based on research supported by the National Science Foundation, the US Department of Energy, and the Army Research Office, most recently by NSF grants OAC-1740309, DMS-1745654, DMS-1906446, Department of Energy grant DE-NA003967, and ARO grant W911NF-16-1-0504. Any opinions, findings, and conclusions or recommendations expressed in this material are those of the author and do not necessarily reflect the views of the National Science Foundation, the Department of Energy, or the Army Research Office.

In addition to the many people who influenced [107] and [111], I want to particularly thank Jerry Bernholc, Wei Bian, Emil Briggs, Luis Chacon, Xiaojun Chen, Kevin Clarno, Austin Ellis, Tom Evans, Elizabeth Greenspan, Steven Hamilton, Michael Herbst, Nick Higham, Ilse Ipsen, Elena Jakubikova, Randy LeVeque, Antoine Levitt, Lin Lin, Wenchang Lu, Juan Meza, Zack Morrow, James Nance, Chung-Wei Ng, Sheehan Olver, Benoit Pasquier, Roger Pawlowski, Liqun Qi, Stuart Slattery, Alex Toth, Romain Veltz, Homer Walker, Jeff Willert, Carol Woodward, Chao Yang, and a few hundred MA 784 students for their inspiration as I got this project into shape.

C. T. Kelley
Raleigh, North Carolina
September 6, 2022

How to Get the Software

This book is tightly coupled to a package of solvers and test problems and a suite of IJulia (Jupyter) notebooks. The print book and the notebooks are very similar, but not identical. The best way to learn this material is to use all three.

The GitHub repositories are

- Notebooks: `https://github.com/ctkelley/NotebookSIAMFANL`
  The notebooks are organized by chapter. `ChapterX.ipynb` is a transcription of the algorithmic discussion in the print book. `ChapterXs.ipynb` is a notebook that lets you play with the examples. That notebook has been mapped to LaTeX and is the notebook section of the corresponding chapter in the print book.

- Julia package: `https://github.com/ctkelley/SIAMFANLEquations.jl`
  The package **SIAMFANLEquations.jl** contains the solvers, the test problems, and the examples. We will refer to the solvers throughout the book. This link is to the main branch, where I'm fixing bugs and updating the documentation.

  When you add the package you will download the stable branch. As I fix bugs, I will update the stable branch.

  The GitHub repositories each have an archival **FA20** branch. The archival branches are the ones that were current when the book was printed. These are

    - `https://github.com/ctkelley/SIAMFANLEquations.jl/tree/FA20`

    - `https://github.com/ctkelley/NotebookSIAMFANL/tree/FA20`

Clone these repositories to have the version which produced the book. The version you get from the package manager when you install the package may have a version number larger that 1.0 as I make corrections or update the codes. I will refer to that version as the stable branch.

Downloads

To get started, download Julia from

`https://julialang.org/downloads/`

When you install Julia it creates a **.julia** subdirectory in your home directory. When you install packages they go in that directory.

You will also put your **startup.jl** file in **.julia/config**.

The solvers and test problems are in a Julia package. You should use the Julia package manager **Pkg** to do this.

- If you only want to use the solvers type

```
import Pkg; Pkg.add("SIAMFANLEquations")
```

  at the ">" prompt in the read-eval-print loop (REPL).

- If you want to play with the source codes, then you want to clone the repository and make any changes you make effective in your Julia environment. In that case, after adding the package, type

```
develop SIAMFANLEquations
```

  at the ">" prompt in the REPL. The **develop** command will clone the repository in your **.julia/dev** directory.

Then type

```
using SIAMFANLEquations
using SIAMFANLEquations.TestProblems
using SIAMFANLEquations.Examples
```

at the REPL prompt. If you intend to make changes to the source codes for the solvers, you should install the **Revise.jl** [92] package so your changes will be recompiled in real time as you work.

While we have tried to minimize dependencies, you will still need several Julia packages to use the solvers and run the test problems and examples. Installing the package

SIAMFANLEquations.jl

should automatically make sure you have the ones for the solvers, examples, and test problems.

- For the solvers:
    - BandedMatrices
    - LaTeXStrings
    - LinearAlgebra
    - Printf
    - SuiteSparse
    - SparseArrays
    - Test

- For the test problems and examples: everything for the solvers and

  – AbstractFFTs

  – FFTW

  – QuadGK

  The case studies in Chapter 5 use both the TestProblems and Examples submodules.

- For the notebooks: everything for the solvers, test problems, and examples plus

  – BenchmarkTools

  – IJulia

  – PyPlot

You should also get the IJulia notebooks from

`https://github.com/ctkelley/NotebookSIAMFANL`

Clone that repository and follow the instructions in the README.md file. You'll need the solvers and the test problems to run the notebook. Once you've installed all the packages then

- Start the first notebook `SIAMFANL.ipynb`.

  – Go to the directory where you've put the notebooks. Start Julia and type

    ```
    notebook(dir=pwd())
    ```

    at the REPL prompt. This will open a webpage with a list of the notebooks. Click on `SIAMFANL.ipynb` or any of the other notebooks.

- The notebook table of contents has links to the chapters.

- When you open any of the notebooks, run all the cells in that notebook. In particular, the first two cells are critical:

  – The first cell in each notebook is a markdown cell with LaTeX definitions. It is not visible, but you need to run it to make the LaTeX in the notebook render correctly.

  – The second cell in the notebook is a code cell with the single command

    ```
    include("fanote_init.jl")
    ```

    This runs a Julia script that organizes the dependencies and functions for the examples.

  You must run these cells whenever you start one of the notebooks.

It's best to install the packages in the Julia REPL. Clicking on the Julia icon will open a terminal window in which you can type commands at the prompt >. To install package **XX.jl** type

```
import Pkg; Pkg.add("XX")
```

at the REPL prompt. It is not a good idea to try to add packages from a notebook code cell. When you install a package you will see several lines of text scroll by during the installation. This is normal.

Once you have installed **XX** you can use the functions inside by typing

```
using XX
```

at the REPL prompt. Then you can get started. For example, to use notebooks you will need to install the **IJulia.jl** package. Having done that, you start a notebook with

```
using IJulia
notebook()
```

The second line opens a browser window with Jupyter ready to go. Then you must navigate to the directory where you put the notebooks. If you are in the directory with the notebooks, then type

```
using IJulia
notebook(;dir=pwd())
```

at the REPL prompt.

Solvers, Test Problems, and Examples

Solvers

The solvers and test problems are part of **SIAMFANLEquations** the Julia package. The solvers are located in **/src/Solvers**.

The core codes and the relevant chapters are

Chap. 1: Solvers for scalar equations **nsolsc.jl**, **secant.jl**, and **ptsolsc.jl** in the **src/Solvers/Chapter1** subdirectory.

Chap. 2: Solvers with Gaussian elimination **nsol.jl** and **ptcsol.jl**.

Chap. 3: Newton–Krylov solvers **nsoli.jl** and **ptcsoli.jl**.

 • GMRES (**kl_gmres.jl**) and BiCGStab (**kl_bicgstab.jl**) linear solvers in the **src/Solvers/LinearSolvers** subdirectory.

Chap. 4: Anderson acceleration **aasol.jl**.

Chap. 5: Case studies: All solvers from Chapters 2, 3, and 4. The case studies are in the **src/TestProblems/CaseStudies** subdirectory.

The Krylov linear solvers **kl_gmres.jl** and **kl_bicgstab.jl** for Chapter 3 can be used for general problems, but have been designed to couple with the nonlinear solvers. In particular, they are aware of any data the function, Jacobian-vector product, or preconditioner may need.

Test Problems and Examples

The test problems and examples are also part of the package because they are used for continuous integration (a.k.a. unit testing). The test problems are in the **/src/TestProblems** directory and the examples in **/src/Examples**. Both the test problems and examples are submodules of **SIAMFANLEquations**. The Julia codes for the figures in the print book and notebooks are in the **Notebook/src** directory.

You can install the solvers without installing the test problems or the examples. To install the test problems and examples type

```
using SIAMFANLEquations.TestProblems
using SIAMFANLEquations.Examples
```

at the REPL prompt. The notebooks use the examples and test problems and the "using" commands are in the file **fanote_init.jl** for that reason.

The notebook knows where the examples and test problems are, so running the examples for yourself should be easy. I hope that changing them and playing with the solvers is equally easy.

Computing Environment for the Book

The results in this book and the docstrings in the codes were obtained on a 2019 Apple iMac with an Intel 8-Core I9 CPU using Julia 1.7.2 and the Mathkernel Library (MKL) Basic Linear Algebra Subprograms (BLAS). The results, especially the tables of errors, will differ from ours depending on your choice of BLAS and the architecture of your computer. In particular, if you use OpenBLAS, the default with Julia, you will see small differences from our results.

The notebooks have been tested with Julia 1.7.2 and 1.6.5. Versions 1.7.0 and 1.6.4 have bugs that cause problems with the notebooks.

Learning Julia

Unfortunately, there is no introduction to the language aimed at the practicing numerical analyst, as there is for MATLAB [89]. There are many online resources for Julia and you should take some care to stick to the ones for version 1.0 and above. We recommend that you use the latest version of Julia.

If you are a Julia novice, it will take some experimentation to get comfortable. One very good reference is the paper [17]. You should read this first. There are also some recent Julia-centric entry-level numerical analysis books [42, 58]. The full manual [188] and the online guide [3] are also very useful. If you are coming from MATLAB, Python, or another language, your best bet is to start off by learning how linear algebra works in Julia. The Wiki [204] is good.

There are many resources at `https://julialang.org/learning`.

- There are general introductions to Julia that are aimed a more general audiences or introductory computer science courses. A good example of such a book is [127], which is also available at

  `https://benlauwens.github.io/ThinkJulia.jl/latest/book.html`

- The Julia YouTube channel

  `https://www.youtube.com/channel/UC9IuUwwE2xdjQUT_LMLONoA`

  has several useful tutorials.

- Two useful and fast introductions are the video

  `https://www.youtube.com/watch?v=8h8rQyEpiZA&t`

  and the collection of webpages [3]

  `https://techytok.com/from-zero-to-julia`

- The discourse at

  `https://discourse.julialang.org`

  is a welcoming environment for the novice.

Of course, the optimal way to learn a computer language is to play with it. The author of this book learned, and is still learning, Julia in that way. Once the reader has a few basic skills, working through this book would be a good way to learn more about the language.

The Repositories

The Julia package and the notebooks are in two GitHub repositories. The repositories are separate because the Julia package manager will put the solver package in the `.julia` subdirectory of your home directory, which is not a convenient place for the notebooks.

Each repository has an archival **FA20** branch which is the one used to produce this book. The master branches are the ones I use for fixing postpublication errors and bugs. The README files on the repositories begin with up-to-date accounts of the status of the package and notebooks.

The README files also tell you which version of Julia created the repository. Your best bet is to use the latest version of Julia. If your organization limits software installation, you may have to use the Long Term Support (LTS) version of Julia, which is currently 1.6.5.

The Package Repository

SIAMFANLEquations.jl is a registered Julia package. The repository has many subdirectories (for example, **/test** and **/docs**) that most readers of this book can ignore.

The **/src** subdirectory has the codes for the solvers, test problems, and examples. I will point the reader to the relevant files for each chapter at the beginning of the chapter.

You should not need to look at the **/src/Tools** directory. The files in that directory manage the internal data structures for the solvers and assemble the output tuples. The only file in **/src/Tools** of algorithmic interest is **armijo.jl**, which manages the line search.

The archival branch at

```
https://github.com/ctkelley/SIAMFANLEquations.jl/tree/FA20
```

contains the files I used to create the book.

The documentation for the stable branch is at

```
https://ctkelley.github.io/SIAMFANLEquations.jl/stable/
```

I built this documentation with the **Documenter.jl** [83] package. It is far less complete than the notebooks. The files that generate the docs are stored in the **/src/docs** directory.

The master (development) branch of the package repository is

```
https://github.com/ctkelley/SIAMFANLEquations.jl
```

I would not advise you to use that branch because I will be changing it as I correct errors and bugs. I will make no changes that alter the user interface to the codes. The branch you get with the Julia package manager is the stable branch. Use that one.

The Notebook Repository

The README file for the notebook repository explains how to set up the notebook and the solver package. The first step is to clone the repository and put the directory in a convenient place.

The root directory contains all the **ipynb** notebook files, an HTML file for the bibliography **siamfa.html**, and a Julia script **fanote_init.jl**. Each notebook runs some LaTeX commands in the first (invisible) cell and **fanote_init.jl** in the second. You should run these two cells before doing anything with a notebook.

Codes for the Examples

The **/src** subdirectory of the Notebook repository has codes that produced the figures and tables in the book. These codes use the examples and test problems from the package.

Notebook Versions

The notebooks have been most recently tested with Julia 1.7.2 and 1.6.5. We recommend that you use the latest version of Julia.

The archival repository is

`https://github.com/ctkelley/NotebookSIAMFANL/tree/FA20`

This is the repository which I used to produce the print book. Versions with numbers larger than 1.0 are for bug fixes and error correction in the text.

The version at

`https://github.com/ctkelley/NotebookSIAMFANL`

is the version I am constantly updating. There should be no harm in cloning that repository and using it.

Chapter 1

Introduction

Files for This Chapter

- From the Package repository:

    - Solvers for scalar equations: **/src/Solvers/Chapter1**

        * Newton's method: **nsolsc.jl**

        * Secant method: **secant.jl**

        * Pseudo-transient continuation: **ptcsolsc.jl**

    - Test problems: **/src/TestProblems/Scalars/spitchfork.jl**

- From the Notebook repository: **/src/Chapter1**
  Julia codes that generate the figures and tables

1.1 ▪ What Is the Problem?

Solving nonlinear equations is part of almost all simulations of physical processes. Physical models that are expressed as nonlinear partial differential equations, for example, become large systems of nonlinear equations when discretized. Developers of simulation codes must either use a nonlinear solver as a tool or write one from scratch. One purpose of this book is to show these developers what technology is available, sketch the implementation, and warn of the problems. We do this via algorithmic outlines, nonlinear solvers in Julia that can be used for production work, a set of example problems, a suite of IJulia (Jupyter) notebooks, a few case studies, and chapter-ending projects.

We use the standard notation

$$\mathbf{F}(\mathbf{x}) = 0 \tag{1.1}$$

for systems of N equations in N unknowns. Here $\mathbf{F} : \Omega \to R^N$, where $\Omega \subset R^N$ is open. We will call $\mathbf{F}$ the **nonlinear residual** or simply the **residual**. Rarely can the solution of a nonlinear equation be given by a closed-form expression, so iterative methods must be used to approximate the solution numerically. The output of an iterative method is a sequence of approximations to a solution.

The **fixed point** formulation of a nonlinear equation is

$$\mathbf{x} = \mathbf{G}(\mathbf{x}). \tag{1.2}$$

The difference between the two formulations is not simply a matter of replacing $\mathbf{F}(\mathbf{x})$ by $\mathbf{x} - \mathbf{G}(\mathbf{x})$. Effective algorithms for (1.1) take a very different approach from those for (1.2), as we will see in Chapter 4.

We will spend most of our time in this introductory section on methods for (1.1). The reason for this is that much of the theory can be explored in the simple context of scalar equations. We will consider fixed point problems seriously in Chapter 4.

1.1.1 ▪ Notation

In this book, following the convention in [107, 110], vectors are to be understood as column vectors. Following [112], we denote vectors by boldfaced lowercase letters and matrices by boldfaced uppercase letters, for example, $\mathbf{x}$ and $\mathbf{A}$. We denote the ith component of $\mathbf{x}$ by x_i to distinguish between the ith member of a sequence of vectors $\mathbf{x}_i$. We denote the ij entry of $\mathbf{A}$ by $\mathbf{A}_{ij}$.

The vector $\mathbf{x}^*$ will denote a solution, $\mathbf{x}$ a potential solution, and $\{\mathbf{x}_n\}_{n \geq 0}$ the sequence of iterates. We will refer to $\mathbf{x}_0$ as the **initial iterate** (not guess!). We will denote the ith component of a vector $\mathbf{x}_n$ from a sequence by x_{ni}. We will rarely need to refer to individual components of vectors. We will let $\partial \mathbf{F} / \partial x_i$ denote the partial derivative of $\mathbf{F}$ with respect to x_i. As is standard [54, 107], $\mathbf{e} = \mathbf{x} - \mathbf{x}^*$ will denote the error. So, for example, $\mathbf{e}_n = \mathbf{x}_n - \mathbf{x}^*$ is the error in the nth iterate.

If the components of $\mathbf{F}$ are differentiable at $\mathbf{x} \in R^N$, we define the **Jacobian matrix** $\mathbf{F}'(\mathbf{x})$ by

$$\mathbf{F}'(\mathbf{x})_{ij} = \frac{\partial f_i}{\partial x_j}(\mathbf{x}).$$

Throughout the book, $\| \cdot \|$ will denote the Euclidean norm on R^N:

$$\|\mathbf{x}\| = \left(\sum_{i=1}^{N} x_i^2 \right)^{1/2}.$$

We treat scalar equations with lowercase letters. So a scalar equation is $f(x) = 0$ and the derivative is $f'(x)$. Many of the essential ideas in this book can be illustrated with scalar equations, and we do that in this chapter. The exception is the linear algebra requirements for solving nonlinear systems of equations. Here we need to solve linear systems of equations and linear least squares problems, which will be the focus of the remaining chapters in the book. The Julia codes for the examples in this chapter are in the **src/Chapter1** directory for the notebook. The solvers and test problems are part of the **SIAMFANLEquations.jl** Julia package.

1.2 ▪ Newton's Method

Most of the methods in this book are variations of Newton's method. The exception will be Anderson acceleration, an algorithm for fixed point problems, which we cover in Chapter 4. The Newton sequence is

$$\mathbf{x}_{n+1} = \mathbf{x}_n - \mathbf{F}'(\mathbf{x}_n)^{-1}\mathbf{F}(\mathbf{x}_n). \tag{1.3}$$

The interpretation of (1.3) is that we model $\mathbf{F}$ at the current iterate $\mathbf{x}_n$ with the linear function

$$\mathbf{M}_n(x) = \mathbf{F}(\mathbf{x}_n) + \mathbf{F}'(\mathbf{x}_n)(\mathbf{x} - \mathbf{x}_n)$$

and let the root of $\mathbf{M}_n$ be the next iteration. $\mathbf{M}_n$ is called the **local linear model**. If $\mathbf{F}'(\mathbf{x}_n)$ is nonsingular, then $\mathbf{M}_n(\mathbf{x}_{n+1}) = 0$ is equivalent to (1.3).

Figure 1.1 illustrates the local linear model and the Newton iteration for the scalar equation

$$\arctan(x) = 0$$

with initial iterate $x_0 = 1$. We graph the local linear model

$$m_j(x) = f(x_j) + f'(x_j)(x - x_j)$$

at x_j from the point $(x_j, y_j) = (x_j, f(x_j))$ to the next iteration $(x_{j+1}, 0)$. The iteration converges rapidly and one can see the linear model becoming more and more accurate. The third iterate is visually indistinguishable from the solution. The Julia program **linear-model.jl** from the Notebook repository created Figure 1.1 by calling the solver **nsolsc.jl** and running some (very messy) **PyPlot** commands. We use the **LaTeXStrings.jl** Julia package [101] in Figure 1.1 and for many other figures in the book.

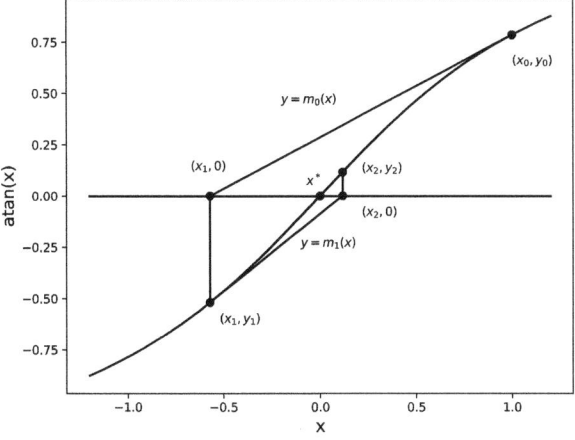

Figure 1.1: Newton iteration for the arctan function.

The computation of a Newton iteration requires

1. evaluation of $\mathbf{F}(\mathbf{x}_n)$ and a test for termination,

2. approximate solution of the equation

$$\mathbf{F}'(\mathbf{x}_n)\mathbf{s} = -\mathbf{F}(\mathbf{x}_n) \tag{1.4}$$

for the Newton step **s**, and

3. construction of $\mathbf{x}_{n+1} = \mathbf{x}_n + \lambda\mathbf{s}$, where the step length λ is selected to guarantee decrease in $\|\mathbf{F}\|$.

Item 2, the computation of the Newton step, consumes most of the work, and the variations in Newton's method that we discuss in this book differ most significantly in how the Newton step is approximated. Computing the step may require evaluation and factorization of the Jacobian matrix or the solution of (1.4) by an iterative method. Not all methods for computing the Newton step require the complete Jacobian matrix, which, as we will see in Chapter 2, can be very expensive.

In the example from Figure 1.1, the step s in item 2 was satisfactory, and we can use $\lambda = 1$ in step 3. The reader should be warned that attention to the step length is generally very important. One should not write one's own nonlinear solver without step-length control (see section 1.6).

1.2.1 ▪ Local Convergence Theory

The convergence theory for Newton's method [54, 107, 149] that is most often seen in an elementary course in numerical methods is **local**. This means that one assumes that the **initial iterate** $\mathbf{x}_0$ is near a solution. The local convergence theory from [54, 107, 149] requires the **standard assumptions**.

Assumption 1.2.1. (standard assumptions)

1. *Equation 1.1 has a solution* $\mathbf{x}^* \in \Omega$.

2. $\mathbf{F}' : \Omega \to R^{N \times N}$ *is Lipschitz continuous near* $\mathbf{x}^*$.

3. $\mathbf{F}'(\mathbf{x}^*)$ *is nonsingular.*

Nonlinear equations can have multiple solutions. The standard assumptions distinguish a single solution $\mathbf{x}^*$ near which the iteration will begin.

Recall that Lipschitz continuity near $\mathbf{x}^*$ means that there is $\gamma > 0$ (the **Lipschitz constant**) such that

$$\|\mathbf{F}'(\mathbf{x}) - \mathbf{F}'(\mathbf{y})\| \le \gamma\|\mathbf{x} - \mathbf{y}\|$$

for all $\mathbf{x}, \mathbf{y}$ sufficiently near $\mathbf{x}^*$.

Theorem 1.1 is the classic local convergence theorem.

Theorem 1.1. *Let the standard assumptions hold. If* $\mathbf{x}_0$ *is sufficiently near* $\mathbf{x}^*$, *then the Newton sequence exists (i.e.,* $\mathbf{F}'(\mathbf{x}_n)$ *is nonsingular for all* $n \ge 0$*) and converges to* $\mathbf{x}^*$ *and there is* $K > 0$ *such that*

$$\|\mathbf{e}_{n+1}\| \le K\|\mathbf{e}_n\|^2 \tag{1.5}$$

for n sufficiently large.

The convergence described by (1.5), in which the norm of the error in the solution will be roughly squared with each iteration, is called **q-quadratic**. Squaring the error roughly means that the number of significant figures in the result doubles with each iteration. Of course, one cannot examine the error without knowing the solution. However, we can observe the quadratic reduction in the error computationally if $\mathbf{F}'(\mathbf{x}*)$ is well-conditioned (see (1.15)), because the norm of the nonlinear residual will also be roughly squared with each iteration. Therefore, we should see the exponent field of the norm of the nonlinear residual roughly double with each iteration.

In Table 1.1 we report the Newton iteration for the scalar ($N = 1$) nonlinear equation

$$f(x) = \tan(x) - x = 0, \, x_0 = 4.5. \tag{1.6}$$

The solution of interest, one of infinitely many, is $x^* \approx 4.493$.

The decrease in the function is as the theory predicts for the first three iterations, then progress slows down for iteration 4 and stops completely after that. The reason for this **stagnation** is clear: one cannot evaluate the function to higher precision than (roughly) machine unit roundoff, which in the IEEE [96,97,150] floating point system is about 10^{-16}.

Table 1.1: Residual history for Newton's method.

n	$|f(x_n)|$
0	1.3733e–01
1	4.1319e–03
2	3.9818e–06
3	5.5955e–12
4	8.8818e–16
5	8.8818e–16

Stagnation is not affected by the accuracy in the derivative. The results reported in Table 1.1 used a forward difference approximation to the derivative with a difference increment of 10^{-7} and report six iterations (0, 1, ..., 5). For this particular example, with this choice of difference increment, the convergence speed of the nonlinear iteration is as fast as that for Newton's method with an analytic derivative until stagnation takes over. The reader should be aware that difference approximations to derivatives, while usually (but not always) reliable, are often expensive. An inaccurate Jacobian can cause many problems (see section 1.9). An analytic Jacobian can require some human effort, but can be worth it in terms of computer time and robustness when a difference Jacobian performs poorly. Automatic differentiation (AD) is also becoming an attractive option [80,94,98,164].

Figure 1.2 is a plot of the progress (or lack of progress) of the iteration for several more iterations and demonstrates that no miracles happen if you keep iterating.

One can quantify this stagnation by adding the errors in the function evaluation and derivative evaluations to Theorem 1.1. Theorem 1.2 sends us the following messages:

- Small errors, for example, machine roundoff, in the function evaluation can lead to stagnation. This type of stagnation is usually benign and, if the Jacobian is well conditioned (see (1.15) in section 1.5), the results will be as accurate as the evaluation of F.

- Small errors in the Jacobian and in the solution of the linear equation for the Newton step (1.4) will affect the speed of the nonlinear iteration, but not the limit of the sequence.

Theorem 1.2. *Let the standard assumptions hold. Let a matrix-valued function $\Delta(\mathbf{x})$ and a vector-valued function $\epsilon(\mathbf{x})$ be such that*

$$\|\Delta(\mathbf{x})\| < \delta_J \text{ and } \|\epsilon(\mathbf{x})\| < \epsilon_F$$

for all $\mathbf{x}$ near $\mathbf{x}^*$. Then, if $\mathbf{x}_0$ is sufficiently near $\mathbf{x}^*$ and δ_J and ϵ_F are sufficiently small, the sequence

$$\mathbf{x}_{n+1} = \mathbf{x}_n - (\mathbf{F}'(\mathbf{x}_n) + \Delta(\mathbf{x}_n))^{-1}(\mathbf{F}(\mathbf{x}_n) + \epsilon(\mathbf{x}_n))$$

is defined (i.e., $\mathbf{F}'(\mathbf{x}_n) + \Delta(\mathbf{x}_n)$ is nonsingular for all n) and satisfies

$$\|\mathbf{e}_{n+1}\| \leq \bar{K}(\|\mathbf{e}_n\|^2 + \|\Delta(\mathbf{x}_n)\|\|\mathbf{e}_n\| + \|\epsilon(\mathbf{x}_n)\|) \tag{1.7}$$

for some $\bar{K} > 0$.

One can ignore the errors in the function in most cases, but one needs to be aware that stagnation of the nonlinear iteration is all but certain in finite precision arithmetic. However, the asymptotic convergence results for exact arithmetic describe the observations well for most problems.

An important application of Theorem 1.2 is the special case where $\delta_J \approx \sqrt{\epsilon_F}$, i.e., the Jacobian error is roughly the same as the square root of the function error. In this case the middle term in the error estimate $\|\Delta(\mathbf{x}_n)\|\|\mathbf{e}_n\|$ is smaller than the maximum of the other two. We state this as a corollary.

Corollary 1.3. *Let the assumptions of Theorem 1.2 hold. Assume that $\delta_J = O(\sqrt{\epsilon_F})$. Then $\mathbf{x}_0$ is sufficiently near $\mathbf{x}^*$:*

$$\|\mathbf{e}_{n+1}\| = O(\|\mathbf{e}_n\|^2 + \epsilon_F). \tag{1.8}$$

There are two examples of particular interest. Suppose ϵ_F is double precision unit roundoff. If one approximates the Jacobian with a forward difference approximation with a difference increment of $O(\sqrt{\epsilon_F})$, then (1.8) holds and the iteration statistics will (usually!—see [120] for an exception) be almost indistinguishable from those with an analytic Jacobian.

Similarly, if one stores the Jacobian in single precision, where the unit roundoff is the square root of ϵ_F, then (1.8) also holds. Our solvers exploit this by (1) using a forward difference Jacobian as the default and (2) allowing a reduced precision Jacobian. The advantage of a forward difference Jacobian is that the user (i.e., you) does not have to invest in the computation of an analytic Jacobian. The disadvantage is that an analytic Jacobian is usually faster, in terms of computer time. The advantages of storing and factoring the Jacobian in single precision [113] are that both the computer time and the storage burden for the linear solve are cut in half. There are very few disadvantages, as (1.8) indicates. We will explore this in detail later in the book.

While Table 1.1 gives a clear picture of quadratic convergence, it's easier to appreciate a graph. Figure 1.2 is a semilog plot of **residual history**, i.e., the norm of the nonlinear residual against the iteration number. The concavity of the plot for the first few iterations is the signature of superlinear convergence. The final two iterations show the onset of stagnation. We used the **semilogy** command from the **PyPlot** package in Julia for this. See the file **newtonstagnation.jl**, which generated Figure 1.2 for an example.

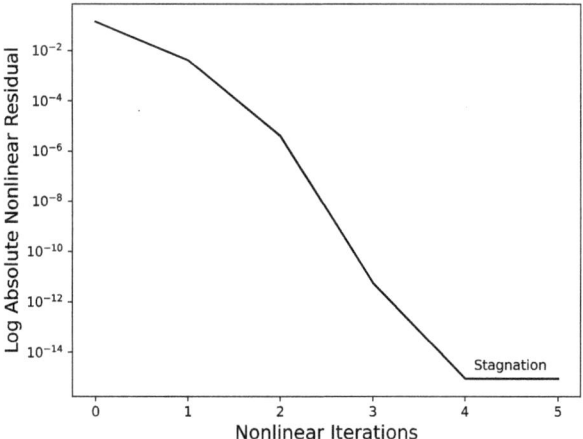

Figure 1.2: Newton iteration for $\tan(x) - x = 0$.

1.3 ▪ Approximating the Jacobian

As we will see in the subsequent chapters, it is usually most efficient to approximate the Newton step in some way. The price for such an approximation is that the nonlinear iteration converges more slowly; i.e., more nonlinear iterations are needed to solve the problem. However, the overall cost of the solve is usually significantly less, because the computation of the Newton step is less expensive.

One way to approximate the Jacobian is to compute $\mathbf{F}'(\mathbf{x}_0)$ and use that as an approximation to $\mathbf{F}'(\mathbf{x}_n)$ throughout the iteration. In this way one amortizes a single evaluation and factorization of $\mathbf{F}'$ over the entire iteration. This is the **chord method** or **modified Newton method**. The convergence of the chord iteration is not as fast as Newton's method. Assuming that the initial iteration is near enough to $\mathbf{x}^*$, the convergence is **q-linear**. This means that there is $\rho \in (0,1)$ such that

$$\|\mathbf{e}_{n+1}\| \le \rho \|\mathbf{e}_n\| \tag{1.9}$$

for n sufficiently large. We can apply Theorem 1.2 to the chord method with $\epsilon_F = 0$ and $\|\Delta(\mathbf{x}_n)\| = O(\|\mathbf{e}_0\|)$ and conclude that ρ is proportional to the initial error. The constant ρ is called the **q-factor**. The formal definition of q-linear convergence allows for faster convergence. q-quadratic convergence is also q-linear, as you can see from the definition (1.5). In many cases of q-linear convergence, one observes that

$$\|\mathbf{e}_{n+1}\| \approx \rho \|\mathbf{e}_n\| \text{ or } \|\mathbf{F}(\mathbf{x}_{n+1})\| \approx \rho \|\mathbf{F}(\mathbf{x}_n)\|.$$

In these cases, q-linear convergence is usually easy to see on a semilog plot of the residual norms against the iteration number. The curve appears to be a line with slope $\approx \log(\rho)$ (can you see why?).

The **secant method** for scalar equations approximates the derivative using a finite difference, but, rather than a forward difference, uses the most recent two iterations to form the difference quotient. So

$$x_{n+1} = x_n - \frac{f(x_n)(x_n - x_{n-1})}{f(x_n) - f(x_{n-1})}, \tag{1.10}$$

where x_n is the current iteration and x_{n-1} is the iteration before that. The secant method must be initialized with two points. One way to do that is to let x_{-1} be a small perturbation of x_0, for example, $x_{-1} = 0.99x_0$. This is what we do in our Julia solver **secant.jl**. The formula for the secant method does not extend to systems of equations ($N > 1$) because the denominator in the fraction would be a difference of vectors. There are many generalizations of the secant method for systems of equations, and we refer the reader to [107] for a discussion of those methods.

The secant method's approximation to $f'(x_n)$ converges to $f'(x^*)$ as the iteration progresses. Theorem 1.2, with $\epsilon = 0$ and $\|\Delta(\mathbf{x}_n)\| = O(\|\mathbf{e}_{n-1}\|)$, implies that the iteration converges **q-superlinearly**. This means that either $\mathbf{x}_n = \mathbf{x}^*$ for some finite n or

$$\lim_{n \to \infty} \frac{\|\mathbf{e}_{n+1}\|}{\|\mathbf{e}_n\|} = 0. \tag{1.11}$$

q-superlinear convergence is hard to distinguish from q-quadratic convergence by visual inspection of the semilog plot of the residual history. The residual curve for q-superlinear convergence is concave down but drops less rapidly than the one for Newton's method.

q-quadratic convergence is a special case of q-superlinear convergence. More generally, if $\mathbf{x}_n \to \mathbf{x}^*$ and, for some $p > 1$,

$$\|\mathbf{e}_{n+1}\| = O(\|\mathbf{e}_n\|^p),$$

we say that $\mathbf{x}_n \to \mathbf{x}^*$ q-superlinearly with **q-order** p.

In Figure 1.3, we compare Newton's method with the chord method and the secant method for our model problem (1.6). We see the convergence behavior that the theory predicts in the linear curve for the chord method and in the concave curves for Newton's method and the secant method. We also see the stagnation in the terminal phase.

The figure does not show the division by zero that halted the secant method computation at iteration 6. The secant method has the dangerous property that the difference between x_n and x_{n-1} could be too small for an accurate difference approximation. The division by zero that we observed is an extreme case.

The Julia codes for our scalar nonlinear solvers are **nsolsc.jl**, which provides Newton and chord solvers and **secant.jl**, the secant method solver. **nsolsc.jl** and **secant.jl** are part of the **SIAMFANLEquations.jl** package and have user interfaces similar to those of the more advanced codes which we cover in subsequent chapters. We will discuss the use of **nsolsc.jl** and **secant.jl** in section 1.10. The code that generated Figure 1.3 is **threewaystagnation.jl** from the notebook.

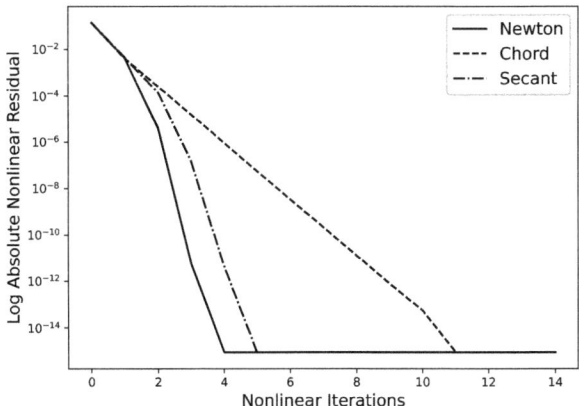

Figure 1.3: Newton/chord/secant comparison for $\tan(x) - x$.

1.4 ▪ Inexact Newton Methods

Approximation of the Jacobian is only one way to solve the equation for the Newton step approximately. An **inexact Newton method** [52] replaces the Newton step with a vector **s** that satisfies the **inexact Newton condition**

$$\|\mathbf{F}'(\mathbf{x}_n)\mathbf{s} + \mathbf{F}(\mathbf{x}_n)\| \leq \eta \|\mathbf{F}(\mathbf{x}_n)\|. \tag{1.12}$$

The parameter η (the **forcing term**) can be varied as the Newton iteration progresses. Choosing a small value of η will make the iteration more like Newton's method, therefore leading to convergence in fewer iterations. However, a small value of η may make computing a step that satisfies (1.12) very expensive. The local convergence theory [52, 107] for inexact Newton methods reflects the intuitive idea that a small value of η leads to fewer iterations. Theorem 1.4 is a typical example of such a convergence result.

Theorem 1.4. *Let the standard assumptions hold. Then there are δ and $\bar{\eta}$ such that if* $\|\mathbf{e}_0\| \leq \delta$ *and* $\{\eta_n\} \subset [0, \bar{\eta}]$, *then the inexact Newton iteration*

$$\mathbf{x}_{n+1} = \mathbf{x}_n + \mathbf{s}_n,$$

where

$$\|\mathbf{F}'(\mathbf{x}_n)\mathbf{s}_n + \mathbf{F}(\mathbf{x}_n)\| \leq \eta_n \|\mathbf{F}(\mathbf{x}_n)\|, \tag{1.13}$$

converges q-linearly to $\mathbf{x}^*$. *Moreover,*

- *if $\eta_n \to 0$, the convergence is q-superlinear, and*

- *if $\eta_n \leq K_\eta \|\mathbf{F}(\mathbf{x}_n)\|^p$ for some $K_\eta > 0$, the convergence is q-superlinear with q-order $1 + p$.*

One can use Theorem 1.4 to analyze the chord method or the secant method. In the case of the chord method, the steps satisfy (1.13) with

$$\eta_n = O(\|\mathbf{e}_0\|),$$

which implies q-linear convergence if $\|\mathbf{e}_0\|$ is sufficiently small. For the secant method, $\eta_n = O(\|\mathbf{e}_{n-1}\|)$, implying q-superlinear convergence.

Theorem 1.4 does not fully describe the performance of inexact methods in practice because the theorem ignores the method used to obtain a step that satisfies (1.12) and ignores the dependence of the cost of computing the step on the forcing term η.

Iterative methods (such as GMRES [170]) for solving the equation for the Newton step would typically use (1.12) as a termination criterion. In this case, the overall nonlinear solver is called a **Newton iterative method**. Newton iterative methods are named by the particular iterative method used for the linear equation. For example, the Newton–Krylov methods in the **nsoli.jl** code, which we describe in Chapter 3, include an implementation of Newton-GMRES.

An unfortunate choice of the forcing term η can lead to very poor results. The reader is invited to try the two choices $\eta = 10^{-6}$ and $\eta = 0.9$ in **nsoli.m** to see this. Better choices of η include $\eta = 0.1$, the author's personal favorite, and a more complex approach (see section 3.4.3) from [60] and [107] that is an option in **nsoli.jl**. Either of these usually leads to rapid convergence near the solution, but at a much lower cost for the linear solver than a very small forcing term such as $\eta = 10^{-6}$.

1.5 ▪ Termination of the Iteration

While one cannot know the error without knowing the solution, in most cases the norm of $\mathbf{F}(\mathbf{x})$ can be used as a reliable indicator of the rate of decay in $\|\mathbf{e}\|$ as the iteration progresses [107]. Based on this heuristic, we terminate the iteration in our codes when

$$\|\mathbf{F}(\mathbf{x})\| \leq \tau_r \|\mathbf{F}(\mathbf{x}_0)\| + \tau_a. \tag{1.14}$$

The relative τ_r and absolute τ_a error tolerances are both important. Using only the relative reduction in the nonlinear residual as a basis for termination (i.e., setting $\tau_a = 0$) is a poor strategy because an initial iterate that is near the solution may make (1.14) impossible to satisfy with $\tau_a = 0$.

One way to quantify the utility of termination when $\|\mathbf{F}(\mathbf{x})\|$ is small is to compare a relative reduction in the norm of the error with a relative reduction in the norm of the nonlinear residual. If the standard assumptions hold and $\mathbf{x}_0$ and $\mathbf{x}$ are sufficiently near the root, then [107]

$$\frac{\|\mathbf{e}\|}{4\|\mathbf{e}_0\|\kappa(\mathbf{F}'(x^*))} \leq \frac{\|\mathbf{F}(\mathbf{x})\|}{\|\mathbf{F}(\mathbf{x}_0)\|} \leq \frac{4\kappa(\mathbf{F}'(x^*))\|\mathbf{e}\|}{\|\mathbf{e}_0\|}, \tag{1.15}$$

where

$$\kappa(\mathbf{F}'(x^*)) = \|\mathbf{F}'(x^*)\|\|\mathbf{F}'(x^*)^{-1}\|$$

is the condition number of $\mathbf{F}'(x^*)$ relative to the norm $\|\cdot\|$. From (1.15) we conclude that if the Jacobian is well-conditioned (i.e., $\kappa(\mathbf{F}'(x^*))$ is not very large), then (1.14) is a useful termination criterion. This is analogous to the linear case, where a small residual implies a small error if the matrix is well-conditioned.

Another approach, which is supported by theory only for superlinearly convergent methods, is to exploit the fast convergence to estimate the error in terms of the step. If the iteration is converging superlinearly, then

$$\mathbf{e}_{n+1} = \mathbf{e}_n + \mathbf{s}_n = o(\|\mathbf{e}_n\|)$$

and hence

$$\mathbf{s}_n = -\mathbf{e}_n + o(\|\mathbf{e}_n\|).$$

Therefore, when the iteration is converging superlinearly, one may use $\|\mathbf{s}_n\|$ as an estimate of $\|\mathbf{e}_n\|$. One can estimate the current rate of convergence from above by

$$\rho_n = \|\mathbf{s}_n\| / \|\mathbf{s}_{n-1}\| \approx \|\mathbf{e}_n\| / \|\mathbf{e}_{n-1}\| \geq \|\mathbf{e}_{n+1}\| / \|\mathbf{e}_n\|.$$

Hence, for n sufficiently large,

$$\|\mathbf{e}_{n+1}\| \leq \rho_n \|\mathbf{e}_n\| \approx \|\mathbf{s}_n\|^2 / \|\mathbf{s}_{n-1}\|.$$

So, for a superlinearly convergent method, terminating the iteration with $\mathbf{x}_{n+1}$ as soon as

$$\|\mathbf{s}_n\|^2 / \|\mathbf{s}_{n-1}\| < \tau \tag{1.16}$$

will imply that $\|\mathbf{e}_{n+1}\| < \tau$.

Termination using (1.16) is only supported by theory for superlinearly convergent methods, but is used for linearly convergent methods in some initial value problem solvers [20, 162]. The trick is to estimate the q-factor ρ, say, by

$$\rho \approx \|\mathbf{s}_n\| / \|\mathbf{s}_{n-1}\| \text{ or } \rho \approx (\|\mathbf{s}_n\| / \|\mathbf{s}_0\|)^{1/n}. \tag{1.17}$$

Assuming that the estimate of ρ is reasonable, then

$$\|\mathbf{e}_n\| - \|\mathbf{s}_n\| \leq \|\mathbf{e}_{n+1}\| \approx \rho \|\mathbf{e}_n\|$$

implies that

$$\|\mathbf{e}_{n+1}\| / \rho \approx \|\mathbf{e}_n\| \leq \|\mathbf{s}_n\| / (1 - \rho). \tag{1.18}$$

Hence, if we terminate the iteration when

$$\|\mathbf{s}_n\| \leq \tau (1 - \rho) / \rho \tag{1.19}$$

and the estimate of ρ is an **overestimate**, then (1.18) will imply that

$$\|\mathbf{e}_{n+1}\| \leq \rho \|\mathbf{s}_n\| / (1 - \rho) \leq \tau.$$

In practice, a safety factor is used on the left side of (1.19) to guard against an underestimate.

If, however, the estimate of ρ is much smaller than the actual q-factor, the iteration can terminate too soon. This can happen in practice if the Jacobian is ill-conditioned and the initial iterate is far from the solution [118].

1.6 ▪ Global Convergence and the Armijo Rule

The requirement in the local convergence theory that the initial iterate be near the solution is more than mathematical pedantry. To see this, we apply Newton's method to find the root $x^* = 0$ of the function $f(x) = \arctan(x)$ with initial iterate $x_0 = 10$. This initial iterate is too far from the root for the local convergence theory to hold. In fact, the step

$$s = \frac{-f(x_0)}{f'(x_0)} \approx \frac{-1.5}{0.01} \approx -150,$$

while in the correct direction, is far too large in magnitude.

The initial iterate and the four subsequent iterates are

$$10, -138, 2.9 \times 10^4, -1.5 \times 10^9, 9.9 \times 10^{17}.$$

As you can see, the Newton step points in the correct direction, i.e., toward $x^* = 0$, but overshoots by larger and larger amounts. The simple artifice of reducing the step by half until $\|\mathbf{F}(\mathbf{x})\|$ has been reduced will usually solve this problem.

In order to clearly describe this, we will now make a distinction between the **Newton direction** $\mathbf{d} = -\mathbf{F}'(\mathbf{x})^{-1}\mathbf{F}(\mathbf{x})$ and the **Newton step** when we discuss global convergence. For the methods in this book, the Newton step will be a positive scalar multiple of the Newton direction. When we talk about local convergence and are taking full steps ($\lambda = 1$ and $\mathbf{s} = \mathbf{d}$), we will not make this distinction and only refer to the step, as we have been doing up to now in this book.

A rigorous convergence analysis requires a bit more detail. We begin by computing the **Newton direction**

$$\mathbf{d} = -\mathbf{F}'(\mathbf{x}_n)^{-1}\mathbf{F}(\mathbf{x}_n).$$

To keep the step from going too far, we find the smallest integer $m \geq 0$ such that

$$\|\mathbf{F}(\mathbf{x}_n + 2^{-m}\mathbf{d})\| < (1 - \alpha 2^{-m})\|\mathbf{F}(\mathbf{x}_n)\| \qquad (1.20)$$

and let the step be $\mathbf{s} = 2^{-m}\mathbf{d}$ and $\mathbf{x}_{n+1} = \mathbf{x}_n + 2^{-m}\mathbf{d}$. The condition in (1.20) is called **sufficient decrease** of $\|\mathbf{F}\|$. The parameter $\alpha \in (0, 1)$ is a small number intended to make (1.20) as easy as possible to satisfy [54, 107, 149]. $\alpha = 10^{-4}$ is typical and used in our codes.

In Figure 1.4, created by **atanarmijo.jl**, we show how this approach, called the **Armijo rule** [9], succeeds. The circled points are iterations for which $m > 0$ and the value of m is above the circle.

Methods like the Armijo rule are called **line search** methods because one searches for a decrease in $\|\mathbf{F}\|$ along the line segment $[\mathbf{x}_n, \mathbf{x}_n + \mathbf{d}]$.

The line search in our codes manages the reduction in the step size with more sophistication than simply halving an unsuccessful step. The motivation for this is that some problems respond well to one or two reductions in the step length by modest amounts (such as 1/2)

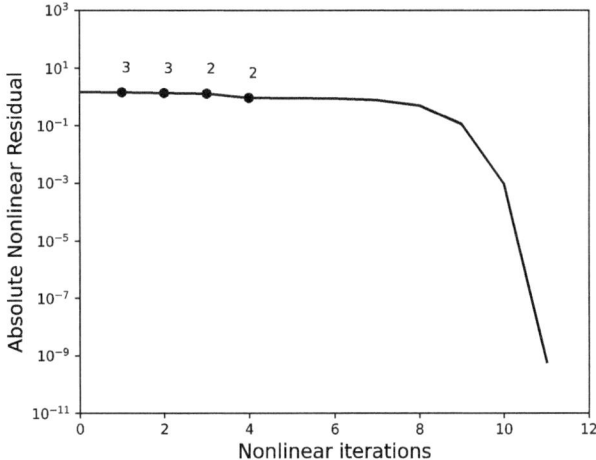

Figure 1.4: Newton–Armijo for $\arctan(x)$.

and others require many such reductions, but might do much better if a more aggressive step-length reduction (by factors of 1/10, say) is used. To address this possibility, after halving the step size fails to not lead to sufficient decrease, we build a quadratic polynomial model of

$$\phi(\lambda) = \|\mathbf{F}(\mathbf{x}_n + \lambda \mathbf{d})\|^2 \tag{1.21}$$

based on interpolation of ϕ at $\lambda = 0$ (the current point) and the two most recent values of λ. The next λ is the minimizer of the quadratic model, subject to the **safeguard** that the reduction in λ be at least a factor of two and at most a factor of ten. So the algorithm generates a sequence of candidate step-length factors $\{\lambda_m\}$ with $\lambda_0 = 0$, $\lambda_1 = 1$ (a full step) and

$$1/10 \le \lambda_{m+1}/\lambda_m \le 1/2. \tag{1.22}$$

The norm in (1.21) is squared to make ϕ a smooth function that can be accurately modeled by a quadratic over small ranges of λ.

The line search terminates with the smallest $m \ge 0$ such that

$$\|\mathbf{F}(\mathbf{x}_n + \lambda_m \mathbf{d})\| < (1 - \alpha \lambda_m)\|\mathbf{F}(\mathbf{x}_n)\|. \tag{1.23}$$

In the advanced codes from the subsequent chapters, we use the three-point parabolic model from [107]. In this approach, $\lambda_1 = 1/2$. To compute λ_m for $m > 1$, a parabola is fitted to the data $\phi(0)$, $\phi(\lambda_m)$, and $\phi(\lambda_{m-1})$. λ_m is the minimum of this parabola on the interval $[\lambda_{m-1}/10, \lambda_{m-1}/2]$. We refer the reader to [107] for the details and to [54, 59, 107, 149] for a discussion of other ways to implement a line search. In Figure 1.5 we apply the parabolic line search to the problem from Figure 1.4. The reader might look at the difference between **atanarmijov2.jl** which makes Figure 1.5 and **atanarmijo.jl** which makes Figure 1.4. As you can see, iteration with the polynomial line search avoids the repeated step size reductions early in the iteration and finds the solution in significantly fewer function evaluations.

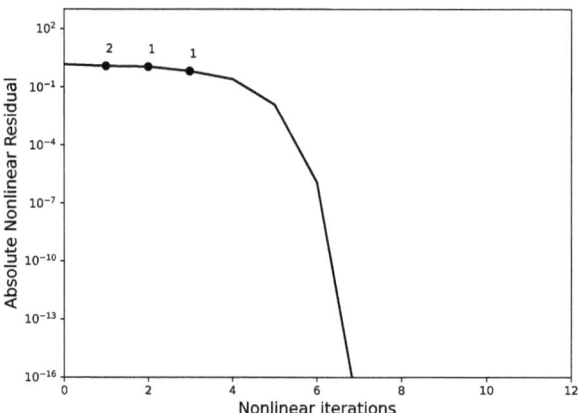

Figure 1.5: Parabolic line search Newton–Armijo for $\arctan(x)$.

1.6.1 ▪ A Basic Algorithm

Algorithm **nsolg** is a general formulation of an inexact Newton–Armijo iteration. The methods in Chapters 2 and 3 are special cases of **nsolg**. There is a lot of freedom in Algorithm **nsolg**. The essential input arguments are the initial iterate **x**, the function **F**, and the relative and absolute termination tolerances τ_a and τ_r. If **nsolg** terminates successfully, **x** will be the approximate solution on output.

Within the algorithm, the computation of the Newton direction **d** can be done with direct or iterative linear solvers, using either the Jacobian $\mathbf{F}'(\mathbf{x})$ or an approximation of it. If you use a direct solver, then the forcing term η is determined implicitly; you do not need to provide one. For example, if you solve the equation for the Newton step with a direct method, then $\eta = 0$ in exact arithmetic. If you use an approximate Jacobian and solve with a direct method, then η is proportional to the error in the Jacobian. Knowing about η helps you understand and apply the theory, but is not necessary in practice if you use direct solvers.

If you use an iterative linear solver, then usually (1.12) is the termination criterion for that linear solver. You'll need to make a decision about the forcing term in that case (or accept the defaults from a code like **nsoli.jl**, which we describe in Chapter 3). The theoretical requirements on the forcing term η are that it be safely bounded away from one (1.24).

Having computed the Newton direction, we compute a step length λ and a step $\mathbf{s} = \lambda \mathbf{d}$ so that the sufficient decrease condition (1.23) holds. It's standard in line search implementations to use a polynomial model like the one we described in section 1.6.

The algorithm does not cover all aspects of a useful implementation. The number of nonlinear iterations, linear iterations, and changes in the step length all should be limited. Failure of any of these loops to terminate reasonably rapidly indicates that something is wrong. We list some of the potential causes of failure in sections 1.9, 2.5, and 3.6.

In [111] the argument list began with **x, F**. Here, following Julia convention [188], we put the function argument first.

ALGORITHM 1.1.
$\mathbf{nsolg}(\mathbf{F}, \mathbf{x}, \tau_a, \tau_r)$

  Evaluate $\mathbf{F}(\mathbf{x})$; $\tau \leftarrow \tau_r \|\mathbf{F}(\mathbf{x})\| + \tau_a$.
  while $\|\mathbf{F}(\mathbf{x})\| > \tau$ **do**
    Find $\mathbf{d}$ such that $\|\mathbf{F}'(\mathbf{x})\mathbf{d} + \mathbf{F}(\mathbf{x})\| \le \eta \|\mathbf{F}(\mathbf{x})\|$.
    If no such $\mathbf{d}$ can be found, terminate with failure.
    $\lambda = 1$
    while $\|\mathbf{F}(\mathbf{x} + \lambda\mathbf{d})\| > (1 - \alpha\lambda)\|\mathbf{F}(\mathbf{x})\|$ **do**
      $\lambda \leftarrow \sigma\lambda$, where $\sigma \in [1/10, 1/2]$ is computed by minimizing the polynomial model
      of $\|\mathbf{F}(\mathbf{x} + \lambda\mathbf{d})\|^2$.
    end while
    $\mathbf{x} \leftarrow \mathbf{x} + \lambda\mathbf{d}$
  end while

The theory for Algorithm **nsolg** is very satisfying. If $\mathbf{F}$ is sufficiently smooth, η is bounded away from one (in the sense of (1.24)), the Jacobians remain well-conditioned throughout the iteration, and the sequence $\{\mathbf{x}_n\}$ remains bounded, then the iteration converges to a solution and, when near the solution, the convergence is as fast as the quality of the linear solver permits. Theorem 1.5 states this precisely, but not as generally as the results in [54, 107, 149]. The important thing that you should remember is that, for smooth $\mathbf{F}$, there are only three possibilities for the iteration of Algorithm **nsolg**:

- $\{\mathbf{x}_n\}$ will converge to a solution $\mathbf{x}^*$, at which the standard assumptions hold,

- $\{\mathbf{x}_n\}$ will be unbounded, or

- $\mathbf{F}'(\mathbf{x}_n)$ will become singular (i.e., $\limsup \|\mathbf{F}'(\mathbf{x}_n)^{-1}\| = \infty$).

Theorem 1.5. *Let $\mathbf{x}_0 \in R^N$ and $\alpha \in (0, 1)$ be given. Assume that $\{\mathbf{x}_n\}$ is given by Algorithm **nsolg**, $\mathbf{F}$ is Lipschitz continuously differentiable,*

$$\{\eta_n\} \subset (0, \bar{\eta}] \subset (0, 1 - \alpha), \tag{1.24}$$

and $\{\mathbf{x}_n\}$ and $\{\|\mathbf{F}'(\mathbf{x}_n)^{-1}\|\}$ are bounded. Then $\{\mathbf{x}_n\}$ converges to a root $\mathbf{x}^*$ of $\mathbf{F}$ at which the standard assumptions hold, full steps ($\lambda = 1$) are taken for n sufficiently large, and the convergence behavior in the final phase of the iteration is that given by the local theory for inexact Newton methods (Theorem 1.4).

While the line search paradigm is the simplest way to find a solution if the initial iterate is far from a root, other methods are available. Trust region globalization [54, 156] is widely used for optimization problems, but less so for nonlinear equations. Pseudo-transient continuation [34, 55, 88, 116] is designed to steer the iteration to dynamically stable solutions. Homotopy methods [16, 203] can be deployed to find all solutions of nonlinear equations. We will cover pseudo-transient continuation in some detail the next section.

1.6.2 ▪ Warning!

The theory for global convergence of the inexact Newton–Armijo iteration is only valid if $\mathbf{F}'(\mathbf{x}_n)$, or a very good approximation (forward difference, for example), is used to compute the step. A poor approximation to the Jacobian will cause the Newton step to be inaccurate. While this can result in slow convergence when the iterations are near the root, the outcome can be much worse when far from a solution. The reason for this is that the success of the line search is very sensitive to the direction. In particular, if $\mathbf{x}_0$ is far from $\mathbf{x}^*$ there is **no reason** to expect the secant or chord method, even with a line search, to converge. Sometimes methods like the secant and chord methods work fine with a line search when the initial iterate is far from a solution, but users of nonlinear solvers should be aware that the line search can fail. A good code will watch for this failure and respond by using a more accurate Jacobian or Jacobian-vector product or reporting an error.

Difference approximations to the Jacobian are usually sufficiently accurate. However, there are particularly hard problems [120] for which differentiation in the coordinate directions is very inaccurate, whereas differentiation in the directions of the iterations, residuals, and steps, which are natural directions for the problem, is very accurate. The inexact Newton methods, such as the Newton–Krylov methods in Chapter 3, use a forward difference approximation for Jacobian-vector products (with vectors that are natural for the problem) and, therefore, will usually (but not always) work well when far from a solution.

1.7 ▪ Pseudo-transient Continuation

Nonlinear equations can have multiple solutions and even if the Newton–Armijo iteration converges to a solution, that may not be the solution you want. One common instance of this problem is integration of an initial value problem to steady state. **Pseudo-transient continuation** (ΨTC) is a method for doing that. Much of this section is taken from [116] and [112]. We refer the reader to [34, 65, 116] for the details of the analysis.

The objective is to find stable steady-state solutions of time-dependent problems

$$\frac{d\mathbf{x}}{dt} = -\mathbf{V}^{-1}\mathbf{F}(\mathbf{x}), \ \mathbf{x}(0) = \mathbf{x}_0. \tag{1.25}$$

In (1.25) the minus sign on the right side of the equation is a convention. The matrix $\mathbf{V}$ is a nonsingular scaling matrix, usually diagonal. $\mathbf{V}$ plays an important role in applications [122], and we include it in the theory, even though we set $\mathbf{V} = \mathbf{I}$ in the examples in this book.

We assume that a solution $\mathbf{x}(t)$ exists for $0 \leq t < \infty$. Here $\mathbf{F}$ is a Lipschitz continuously differentiable function in a neighborhood of the trajectory $\{\mathbf{x}(t) \mid 0 \leq t \leq \infty\}$. We assume that a **steady-state** solution $\mathbf{x}^*$ of (1.25) exists. This means that

$$\lim_{t \to \infty} \mathbf{x}(t) = \mathbf{x}^*. \tag{1.26}$$

Equation (1.26) implies that

$$\frac{d\mathbf{x}^*}{dt} = 0 = -\mathbf{F}(\mathbf{x}^*).$$

A steady state solution is **stable** if the solution of the initial value problem (1.25) with initial data sufficiently near $\mathbf{x}^*$ converges to $\mathbf{x}^*$ as $t \to \infty$. We will only consider linear stability, which for us will mean that there is $\beta > 0$ such that

$$\|(\mathbf{I} + \delta\mathbf{V}^{-1}\mathbf{F}'(\mathbf{x}^*))^{-1}\| \leq (1 + \beta\delta)^{-1}$$

for all $\delta > 0$. This will imply stability, but the converse is not true.

1.7.1 ▪ An ODE Example

A simple ordinary differential equation makes the point. In this example $f(x) = (x^3 - \lambda x)$, where λ is a parameter in the equation. The equation

$$\frac{dx}{dt} = -(x^3 - \lambda x) \qquad (1.27)$$

has a unique steady-state solution $x \equiv 0$ if $\lambda \leq 0$. That solution is stable if $\lambda < 0$ because $f'(0) = -\lambda > 0$. When $\lambda > 0$, however, there are three steady-state solutions

$$x \equiv \pm\sqrt{\lambda} \quad \text{and} \quad x \equiv 0.$$

The two nonzero solutions are stable and $x \equiv 0$ is not. If one solves $f(x) = 0$ with Newton's method, the iteration is

$$x_+ = \frac{-2x_c^3}{\lambda - 3x_c^2}.$$

Hence Newton's method will converge to the unstable solution if the initial iterate x_0 is sufficiently small.

The solution of the initial value problem, on the other hand, will converge to one of the steady-state solutions. Similarly, numerical integration with Euler's method with a time step δ_t,

$$x_{k+1} = x_k - \delta_t f(x_k) = x_k - \delta_t(x_k^3 - \lambda x_k) = (1 + \delta_t\lambda)x_k - \delta_t x_k^3,$$

will converge as $k \to \infty$ to a stable steady-state solution solution if $x_0 \neq 0$ and δ_t is sufficiently small. To see that, note that $x_{k+1} > x_k$ if $x_k > 0$ is small. Hence the numerical integration converges to the stable steady-state solution for sufficiently small δ_t. So the problem is that Newton's method can converge to the unstable solution, which is not of interest, but a highly accurate temporal integration, while finding a steady-state solution, is very costly.

1.7.2 ▪ The ΨTC Algorithm

The ΨTC sequence $\{\mathbf{x}_n\}$ is

$$\mathbf{x}_{n+1} = \mathbf{x}_n - (\delta_n^{-1}\mathbf{V} + \mathbf{F}'(\mathbf{x}_n))^{-1}\mathbf{F}(\mathbf{x}_n). \qquad (1.28)$$

If δ_n were small and fixed, this would be a Rosenbrock method [72] for temporal integration. The objective of ΨTC is fast convergence to steady state. That is very different from accurate temporal integration, hence the name **pseudo-transient** continuation. Another way to view (1.28) is as a single Newton step for the implicit Euler time step

$$\mathbf{x}_{n+1} = \mathbf{x}_n + \delta_n\mathbf{V}^{-1}\mathbf{F}(\mathbf{x}_{n+1}).$$

This viewpoint plays an important role in the theory from [34, 35, 65, 116]. We think of ΨTC as a variable time step algorithm that attempts to increase the time step as $\|\mathbf{F}(\mathbf{x}(t))\|$ becomes small. So we manage δ very differently from a variable-step initial value problem solver, where time step control is for stability and temporal accuracy.

We will consider the simplest case for smooth $\mathbf{F}$ and ordinary differential equation dynamics. ΨTC has also succeeded with differential algebraic equations and nonsmooth dynamics [34, 35, 65].

The formal statement of the algorithm is a bit simpler than that for Newton–Armijo. The implementation is also simpler. For Newton–Armijo, as we shall see, one has options in the frequency of the Jacobian evaluation and factorization, the nature of any approximation of the Jacobian, and the rules for stepsize control. ΨTC , on the other hand, needs to use the Jacobian at the current iteration to preserve the dynamics early in the iteration and controls δ with a simple formula.

We update δ with the "switched evolution relaxation" (SER) method [140], an approach which is widely used in aerodynamics [121, 147, 148, 199]:

$$\delta_n = \delta_{n-1} \|\mathbf{F}(\mathbf{x}_{n-1})\| / \|\mathbf{F}(\mathbf{x}_n)\| = \delta_0 \|\mathbf{F}(\mathbf{x}_0)\| / \|\mathbf{F}(\mathbf{x}_n)\| \tag{1.29}$$

or

$$\delta_n = \phi\left(\delta_{n-1} \frac{\|\mathbf{F}(\mathbf{x}_{n-1})\|}{\|\mathbf{F}(\mathbf{x}_n)\|}\right). \tag{1.30}$$

In (1.30)

$$\phi(\xi) = \begin{cases} \xi, & \xi \le \xi_t, \\ \delta_{\max}, & \xi > \xi_t, \end{cases} \tag{1.31}$$

where either $\xi_t = \delta_{\max}$ or $\xi_t < \infty$ and $\delta_{\max} = \infty$. The choice $\delta_{\max} = \infty$ in (1.30) is (1.29).

Algorithm **ptc** has been used in aerodynamics [199], hydrology [64], mechanics [73], magnetohydrodynamics [123], radiation transport [178], reacting flow [183], structural analysis [103], optimization [117], and circuit simulation [78]. In all of these applications, time-accurate integration to steady state is far too costly to be useful.

ALGORITHM 1.2.
ptc$(\mathbf{F}, \mathbf{x}, \mathbf{V}, \delta, \tau_a, \tau_r)$
    Evaluate $\mathbf{F}(\mathbf{x})$; $\tau \leftarrow \tau_r \|\mathbf{F}(\mathbf{x})\| + \tau_a$.
    while $\|\mathbf{F}(\mathbf{x})\| > \tau$ **do**
        Find $\mathbf{s}$ such that $\|(\delta^{-1}\mathbf{V} + \mathbf{F}'(\mathbf{x}))\mathbf{s} + \mathbf{F}(\mathbf{x})\| \le \eta \|\mathbf{F}(\mathbf{x})\|$.
        If no such $\mathbf{s}$ can be found, terminate with failure.
        $\mathbf{x} \leftarrow \mathbf{x} + \mathbf{s}$
        Evaluate $\mathbf{F}(\mathbf{x})$; store $\|\mathbf{F}(\mathbf{x})\|$.
        Update δ with SER.
    end while

Unlike the Newton–Armijo algorithm, ΨTC allows $\|\mathbf{F}(\mathbf{x})\|$ to increase as the iteration progresses. ΨTC responds to such an increase by decreasing δ and thereby better following transient behavior. This is exactly the right thing to do. Insisting on a decrease in $\|\mathbf{F}(\mathbf{x})\|$, as Newton–Armijo does, can lead to convergence to an unstable steady state or even stagnation at local minimum of $\|\mathbf{F}\|$ [122], at which $\mathbf{F}'$ is singular. These problems are common when the time-dependent solution has complex features such as shocks and rarefactions on the trajectory between $\mathbf{x}_0$ and $\mathbf{x}^*$ (see [147, 148], for example). ΨTC succeeds in many of these cases by taking advantage of the dynamic structure of the problem and following the dynamics early in the iteration.

1.7.3 ▪ Convergence Theory

We begin by summarizing the assumptions from [116] we need for convergence.

Assumption 1.7.1.

- *The initial value problem* (1.25) *has a steady-state solution* $\mathbf{x}^* = \lim_{t \to \infty} \mathbf{x}(t)$.

- *There is a neighborhood of the trajectory* $\{\mathbf{x}(t) \mid 0 \leq t < \infty\}$ *in which* $\mathbf{F}'$ *is uniformly Lipschitz continuous.*

- *There is* $\beta > 0$ *such that*

$$\|(\mathbf{I} + \delta \mathbf{V}^{-1} \mathbf{F}'(\mathbf{x}^*))^{-1}\| \leq (1 + \beta \delta)^{-1}$$

for all $\delta > 0$.

The convergence result is very different from Theorem 1.5. In Theorem 1.6 we specify the target solution $\mathbf{x}^*$ in advance and link $\mathbf{x}_0$ to $\mathbf{x}^*$ with the dynamics. The dynamic stability assumption on $\mathbf{F}(\mathbf{x}^*)$ is stronger than in the standard assumptions for local convergence as well.

Theorem 1.6. *Let Assumption 1.7.1 hold and assume that* δ *is updated by SER. Let* $\{\mathbf{x}_n\}$ *be the iteration from Algorithm* **ptc**. *Assume that* $\eta_n \leq \bar{\eta}$ *for all n. Then if* δ_0 *and* $\bar{\eta}$ *are sufficiently small, then* $\mathbf{x}_n \to \mathbf{x}^*$ *and* $\delta_n \to \delta_{max}$. *Moreover, for n sufficiently large,*

$$\|\mathbf{e}_{n+1}\| = O\left((\eta_n + \delta_n^{-1})\|\mathbf{e}_n\| + \|\mathbf{e}_n\|^2\right). \tag{1.32}$$

1.7.4 ▪ Computational Example

We illustrate the effects of ΨTC with the example from section 1.7.1. We compare a Newton iteration to a ΨTC iteration for $\lambda = 0.5$ with $x_0 = 0.1$. ΨTC will move away from the unstable solution $x_{us}^* = 0$ and converge to the stable solution $x^* = \sqrt{2}/2$.

The reader must be aware that ΨTC **is NOT a general-purpose nonlinear solver**. We will return to this theme in Chapters 2 and 3. ΨTC was invented to find steady-state solutions of time-dependent problems. The price for that ability is that ΨTC generally needs many more iterations to do its work and must (in both theory and practice) use an accurate Jacobian, not, for example, the Jacobian from a prior iteration.

Figure 1.6 compares the convergence of ΨTC and Newton's method on the example problem $f(x) = (x^3 - \lambda x)$ from section 1.7.1. Here we take $\lambda = 1/2$ and $x_0 = 0.1$. The ΨTC iteration will converge to the positive steady-state solution $x^* = \sqrt{1/2}$, while Newton's method converges to the unstable solution $x^* = 0$. We plot both the residual norm and the error. There are two things to mention in this plot.

- ΨTC takes many more iterations to converge. This is because it accurately tracks the dynamics early in the iteration and takes several steps as it does that.

- The two plots look very similar because the derivative of f at either of the two solutions is neither near zero nor too large.

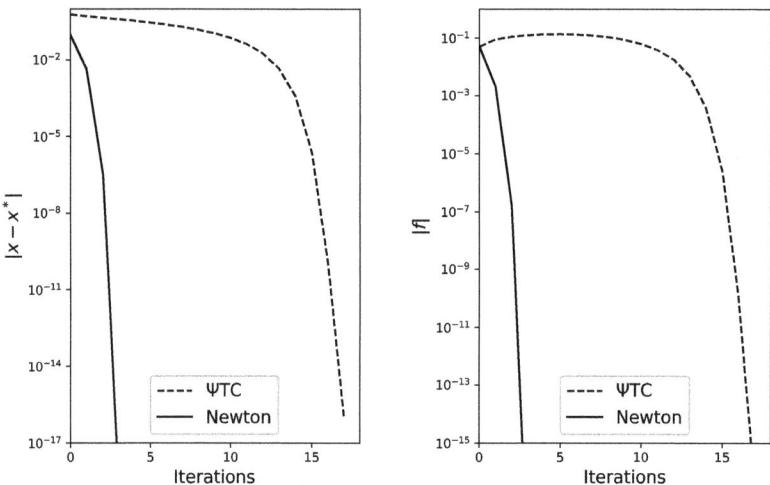

Figure 1.6: ΨTC and Newton iterations.

Figure 1.7 shows how the choice of δ_0 affects the convergence. In this example reducing δ_0 forces a more accurate and costly resolution of the transients in the dynamics. Since our choice of $\delta_0 = 0.1$ found the correct steady-state solution, smaller values of δ_0 only result in wasted effort. Of course, a value of δ_0 that is too large will lead to convergence to the unstable solution, as we will see in section 1.10.3.

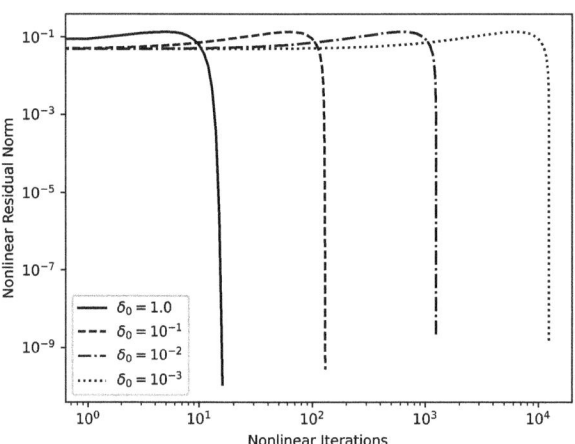

Figure 1.7: Effects of δ_0.

1.8 ▪ Things to Consider

Here is a short list of things to think about when you select and use a nonlinear solver.

1.8.1 ▪ Human Time and Public Domain Codes

When you select a nonlinear solver for your problem, you need to consider not only the computational cost (in time and storage) but also **your time**. A fast code for your problem that takes ten years to write has little value.

Unless your problem is very simple, or you're an expert in this field, your best bet is to use an open source or public domain code. The Julia codes that accompany this book are a good start and can be used for small- to medium-scale production work. The MATLAB codes from [107, 111] serve the same purpose. However, if you need support for other languages (meaning C, C++, or FORTRAN) or leadership-class computing environments at the national laboratories, there are several sources for public domain implementations of the algorithms in this book.

The Newton–Krylov solvers we discuss in Chapter 3 are at present (2022) the solvers of choice for most large problems on advanced computers. Therefore, these algorithms are getting most of the attention from the people who build libraries. The NOX solver in the Trilinos framework [86], the SNES solver in the PETSc library [12, 13], and the NITSOL [151], NKSOL [26], and KINSOL [38, 187] codes are good implementations. KINSOL has an interface to Julia as part of the **SUNDIALS.jl** package [91, 161].

The methods from Chapter 2, which are based on direct factorizations, have received less attention recently and many of the codes are quite old. Some careful implementations can be found in the MINPACK and UNCMIN libraries. The MINPACK [139] library is a suite of FORTRAN codes that includes an implementation of Newton's method for dense Jacobians. The globalization is via a trust region approach [54, 156] rather than the line search method we use here. The UNCMIN [173] library is based on the algorithms from [54] and includes a Newton–Armijo nonlinear equations solver. MINPACK and several other codes for solving nonlinear equations are available from the NETLIB repository at `http://www.netlib.org/`. NETLIB codes are quite old and may not be maintained.

The solvers in **SIAMFANLEquations.jl** were designed to support this book and help the reader understand how to use the algorithms and explore the options. This is a very different mission from that of the other nonlinear solver packages in Julia. Three of these packages are **Sundials.jl**, **BifurcationKit.jl**, and **NLsolve.jl**. At the high end, KINSOL is part of **Sundials.jl**, the Julia interface to Sundials [91, 161] (see `https://github.com/SciML/Sundials.jl`). This is a well-done and important project, but not one designed for a novice to understand. The pathfollowing and bifurcation analysis package **BifurcationKit.jl** [198] has special-purpose nonlinear solvers that communicate well with continuation algorithms. Solvers like **BifurcationKit.jl** and **NLsolve.jl** [137] are highly abstracted. The Julia ecosystem has many codes like this for all kinds of things. They are very useful but hard to learn from. **NLsolve.jl** seems to be based on [54].

Jacobians and even directional derivatives are difficult or impossible to obtain in some applications. The SCF (self-consistent field) iteration in electronic structure computations (see [50, 128, 169] and the references in these papers) in computational physics and chemistry is one such example. Anderson acceleration [6], which we cover in Chapter 4, is one approach for such problems.

There are implementations of Broyden's method, the natural generalization of the secant method to nonlinear systems, in many libraries. The modern implementation in Trilinos is based on [107]. Quasi-Newton methods like Broyden's method have been mostly replaced by Newton–Krylov methods in applications, and we do not cover them in this book.

1.8.2 ▪ The Initial Iterate

Picking an initial iterate at random or without thinking (the famous "initial guess") is a bad idea. Some problems come with a good initial iterate. However, it is usually your job to create one that has as many properties of the solution as possible. Thinking about the problem and the qualitative properties of the solution while choosing the initial iterate can ensure that the solver converges more rapidly and avoids solutions that are not the ones you want.

In some applications the initial iterate is known to be good, so methods like the chord method become very attractive, since the problems with the line search discussed in section 1.6.2 are not an issue. Two examples of this are implicit methods for temporal integration (see section 2.7.4), in which the initial iterate is the output of a predictor, and **nested iteration** (see section 2.9.3), where problems such as differential equations are solved on a coarse mesh and the initial iterate for the solution on finer meshes is an interpolation of the solution from a coarser mesh.

It is more common to have a little information about the solution in advance, in which case one should try to exploit those data about the solution. For example, if your problem is a discretized differential equation, make sure that any boundary conditions are reflected in your initial iterate. If you know the signs of some components of the solution, be sure that the signs of the corresponding components of the initial iterate agree with those of the solution.

1.8.3 ▪ Computing the Newton Step

If function and Jacobian evaluations are very costly, the Newton–Krylov methods from Chapter 3 are worth exploring. These methods avoid explicit computation of Jacobians, but usually require preconditioning (see sections 3.3, 3.4.2, and 4.1.2).

For very large problems, storing a Jacobian is difficult and factoring one may be impossible. Low-storage Newton–Krylov methods, such as Newton-BiCGSTAB, may be the only choice. Even if the storage is available, factorization of the Jacobian is usually a poor choice for very large problems, so it is worth considerable effort to build a good preconditioner for an iterative method. If these efforts fail and the linear iteration fails to converge, then you must either reformulate the problem or find the storage for a direct method.

A direct method is not always the best choice for a small problem, though. Integral equations, such as the example in sections 2.7.2 and 3.8.2, are one type for which iterative methods perform better than direct methods even for problems with small numbers of unknowns and dense Jacobians.

1.8.4 ▪ Choosing a Solver

The most important issues in selecting a solver are

- the size of the problem,

- if derivative information is possible to get or approximate,

- the cost of evaluating $\mathbf{F}$ and $\mathbf{F}'$, and

- the way linear systems of equations will be solved.

The items in the list above are not independent.

The reader in a hurry could use the outline below and probably do well.

- If N is small and evaluation of $\mathbf{F}$ is cheap, computing $\mathbf{F}'$ with forward differences and using direct solvers for linear algebra makes sense. The methods from Chapter 2 are a good choice. These methods are probably the optimal choice in terms of saving your time.

- If the Jacobian is sparse and you can compute it by hand, then storing it in sparse matrix format is all you need to do to inform **nsol.jl** that it should use sparse linear solvers. You should also exploit any special structure you can. For example, if the Jacobian is banded with a small bandwidth, you should exploit that and store the Jacobian as a banded matrix. We use the package **BandedMatrices.jl** [144] in an example in section 2.7.3.

  Sparse differencing can be done in considerable generality [37, 41]. If you can exploit sparsity in the Jacobian, you will save a significant amount of work in the computation of the Jacobian and may be able to use a direct solver. We do not include sparse differencing in our solvers because we believe that is best left to the user and we want to limit the dependencies in the **SIAMFANLEquations.jl** package. The **SparseDiffTools.jl** [160] package is a comprehensive suite of sparse differencing tools.

  You should know that a sparse factorization may have a large storage cost and that allocating that memory, especially in Julia, can be very costly. If you can obtain the sparsity pattern easily, preallocating that memory is a very good idea, and our solvers insist on that. The **SparseDiffTools.jl** can help you do that if you do not know the sparsity pattern. Direct sparse solvers are also very efficient. The Julia package **SuiteSparse.jl** uses the codes from [44, 45]. We will say more about using these methods in Chapter 2.

- If N is large or computing and storing $\mathbf{F}'$ is very expensive, you may not be able to use a direct method.

  - If you can't compute or store $\mathbf{F}'$ at all, but can compute or approximate matrix-vector products, then the matrix-free Newton–Krylov methods in Chapters 3 and Anderson acceleration from Chapter 4 may be your best options. If you have a good preconditioner, a Newton–Krylov code is a good start. The discussion in section 3.2 will help you choose a Krylov method.

    – If $\mathbf{F}'$ is sparse, but you are not able to store the sparse factorization, you may still be able to exploit that structure with an incomplete factorization [167] preconditioner or an in-place solve which may require significantly less storage.

    – If computing derivative information is too costly or impossible, then consider Anderson acceleration (see Chapter 4). While Newton–Krylov should be more efficient if you have a good preconditioner [82], there are important applications where Newton–Krylov methods are not practical.

1.9 ▪ What Can Go Wrong?

Even the best and most robust codes can (and do) fail in practice. In this section we give some guidance that may help you troubleshoot your own solvers or interpret hard-to-understand results from solvers written by others. These are some problems that can arise for all choices of methods. We will also repeat some of these things in subsequent chapters, when we discuss problems that are specific to a method for approximating the Newton direction.

1.9.1 ▪ Nonsmooth Functions

Most nonlinear equation codes, including the ones that accompany this book, are intended to solve problems for which $\mathbf{F}'$ is Lipschitz continuous. While there is room for continuous but not Lipschitz continuous Jacobians [105], the codes can behave unpredictably if your function is not Lipschitz continuously differentiable. If, for example, the code for your function contains

- nondifferentiable functions such as the absolute value, a vector norm, or a fractional power,

- internal interpolations from tabulated data,

- randomized algorithms, such as Monte Carlo simulations,

- control structures like *case* or *if-then-else* that govern the value returned by $\mathbf{F}$, or

- calls to other codes,

you may well have a nondifferentiable problem.

If your function is close to a smooth function, the codes in **SIAMFANLEquations.jl** may do very well. On the other hand, a nonsmooth nonlinearity can cause any of the failures listed in this section.

There are generalizations of Newton's method for broad classes of nonsmooth problems [158], but those methods, while very much worth learning, are far beyond the scope of this book.

1.9.2 ▪ Failure to Converge

The theory for Newton's method, as stated in Theorem 1.5, does not imply that the iteration will converge, only that nonconvergence can be identified easily. So, if the iteration fails to converge to a root, then either the iteration will become unbounded or the Jacobian will become singular. Pseudo-transient continuation and Anderson acceleration can also fail to converge. When this happens, you may have to reformulate your problem, add more/better physics, and think about the units you're using.

Inaccurate function evaluation

Most nonlinear solvers, including the ones that accompany this book, assume that the errors in the evaluation are on the order of machine roundoff and, therefore, use a difference increment of $dx \approx 10^{-7}$ for finite difference Jacobians and Jacobian-vector products. If the error in your function evaluation is larger than that, the Newton direction can be poor enough for the iteration to fail. Thinking about the errors in your function and, if necessary, changing the difference increment in the solvers will usually solve this problem. If your errors are in the formulation rather than in the computation of the function, you may well happily converge to an incorrect result. **nsol.jl** lets you change the difference increment dx if the error in **F** is large.

No solution

If your problem has no solution, then any solver will have trouble. The clear symptoms of this are divergence of the iteration to infinity or failure of the residual to converge to zero. The causes in practice are less clear; errors in programming (a.k.a. bugs) are the likely source. If **F** is a model of a physical problem, the model itself may be wrong. The algorithm for computing **F**, while technically correct, may have been realized in a way that destroys the solution. For example, internal tolerances to algorithms within the computation of **F** may be too loose, internal calculations based on table lookup and interpolation may be inaccurate, and if-then-else constructs can make **F** nondifferentiable.

For example, if $f(x) = e^{-x}$, then the Newton iteration will diverge to $+\infty$ from any starting point. If $f(x) = x^2 + 1$, the Newton–Armijo iteration will converge to 0, the minimum of $f(x)$, which is not a root.

Singular Jacobian

The case where **F**' approaches singularity is particularly dangerous. In this case the step lengths approach zero, so if one terminates when the step is small and fails to check that **F** is approaching zero, one can incorrectly conclude that a root has been found. The example in section 2.7.1 illustrates how an unfortunate choice of initial iterate can lead to this behavior.

If $\mathbf{F}'(\mathbf{x}^*)$ is singular the local convergence theory does not hold. However, the iteration may still converge [48], but not superlinearly. The equation $f(x) = x^2 = 0$ is a good example of this. In this example $x^* = 0$ and for any $x_c \neq 0$, $x_+ = x_c/2$. Hence, if the initial iterate is nonzero, the iteration will converge with q-factor $1/2$.

Alternatives to Newton–Armijo

If you find that a Newton–Armijo code fails for your problem, there are alternatives to line search globalization that, while complex and often more costly, can be more robust than Newton–Armijo. Among these are trust region [54, 156] and homotopy [16, 19, 203] methods. The field of computational algebraic geometry seeks to find all solutions of a polynomial system. There is software linked to the recent book [16] and a Julia package

[19] **HomotopyContinuation.jl**. Pseudo-transient continuation [116] (see sections 1.7, 2.7.6, and 3.5) is also an alternative and we provide codes for that. If these methods fail, you should see if you've made a modeling error and thus posed a problem with no solution.

1.9.3 ▪ Failure of the Line Search

If the line search reduces the step size to an unacceptably small value and the Jacobian is not becoming singular, then the quality of the Newton direction is poor. We repeat the caution from section 1.6.2 that the theory for convergence of the Armijo rule depends on using an accurate Jacobian (analytic or forward difference). A difference approximation to a Jacobian or Jacobian-vector product is usually, but not always, sufficient.

The difference increment in a forward difference approximation to a Jacobian or a Jacobian-vector product should be a bit more than the square root of the error in the function. Our codes use $dx = 10^{-7}$ as the default. This is a good choice unless the function contains components such as a table lookup or output from an instrument that would reduce the accuracy. Central difference approximations, where the optimal increment is roughly the cube root of the error in the function, might (rarely) improve the performance of the solver, but for large problems the cost—twice that of a forward difference—is rarely justified. If you're using a direct method to compute the Newton step, an analytic Jacobian may make the line search perform better.

Finally, one should **scale** the finite difference increment to reflect the size of **x** (see section 2.3).

Failure of the line search in a Newton–Krylov iteration may be a symptom of loss of orthogonality in the linear solver. See section 3.6.2 for more about this problem.

1.9.4 ▪ Slow Convergence

If you use Newton's method and observe slow convergence, the chances are good that the Jacobian, Jacobian-vector product, or linear solver is inaccurate. The local superlinear convergence results from Theorems 1.1 and 1.4 only hold if the correct linear system is solved to high accuracy.

If you expect to see superlinear convergence, but do not, you might consider these things:

- If the errors in **F** are significantly larger than floating point roundoff, then increase the difference increment in a difference Jacobian to roughly the square root of the errors in the function [107].

- Check your computation of the Jacobian (by comparing it to a forward difference approximation, for example).

- If you are using a sparse-matrix code to solve for the Newton step, be sure that you have specified the correct sparsity pattern.

- Make sure the tolerances for an iterative linear solver are set tightly enough to get the convergence you want. Check for errors in the preconditioner and try to investigate its quality.

- If you are using a GMRES solver, make sure that you have not lost orthogonality (see section 3.6.2).

- Do the standard assumptions hold? In particular, is $\mathbf{F}'(\mathbf{x}^*)$ singular?

1.9.5 ▪ Multiple Solutions

In general, there is no guarantee that a nonlinear equation has a unique solution. The solvers we discuss in this book, as well as the alternatives we listed in section 1.9.2, are supported by theory that says that either the solver will converge to a root or it will fail in some well-defined manner. No theory can say that the iteration will converge to the solution that you want. The problems we discuss in sections 2.7.2, 2.7.3, 1.7, 2.7.6, and 3.5 have multiple solutions. In the case of ΨTC , a choice of δ_0 that is too large can lead to convergence to an unstable solution or a stable solution other than the one you want. We discuss this issue in more detail in sections 1.10.3 and 2.7.6.

1.9.6 ▪ Storage Problems

If your problem is large and the Jacobian is dense, you may be unable to store that Jacobian. If your Jacobian is sparse, you may not be able to store the factors that the sparse Gaussian elimination in Julia creates. Even if you use an iterative method, you may not be able to store the data that the method needs to converge. GMRES needs a vector for each linear iteration, for example. Many computing environments, Julia among them, will tell you that there is not enough storage for your job. Julia, for example, will print this message:

```
ERROR: OutOfMemoryError()
```

When this happens, you need to find a way to obtain more memory or a larger computer, or use a solver that requires less storage. Low-storage Newton–Krylov methods (such as Newton-BiCGSTAB and Newton-GMRES(m)) and Anderson acceleration are good candidates.

Other computing environments solve runtime storage problems with virtual memory. This means that data are sent to and from disk as the computation proceeds. This is called **paging** and will slow down the computation by factors of 100 or more. This is rarely acceptable. Your best option is to find a computer with more memory.

Modern computer architectures have complex memory hierarchies. The registers in the CPU are the fastest, so you do best if you can keep data in registers as long as possible. Below the registers can be several layers of cache memory. Below the cache is RAM, and below that is disk. Cache memory is faster than RAM, but much more expensive, so a cache is small. Simple things such as ordering loops to improve the locality of reference can speed up a code dramatically. Thinking about this is a good idea in Julia, as it is in FORTRAN or C. The discussion of loop ordering in [53] is a good place to start learning about efficient programming for computers with memory hierarchies.

Memory allocation is expensive in most computing environments and Julia is particularly sensitive to this. Throughout the book we will point out ways to avoid allocations. The more mature reader may remember having to do this in FORTRAN years ago [57]. The world has not changed much and the parallels with Julia's use of LAPACK [7] and the way people did things 40 years ago are remarkable.

1.10 ▪ Notebook: Codes for Scalar Equations

We provide three solvers for this chapter. **nsolsc.jl** is a scalar Newton code and **ptcsolsc.jl** is a scalar pseudo-transient continuation solver. These scalar codes are simply wrappers for the general codes **nsol.jl** and **ptcsol.jl**. We also include **secant.jl**, which is the secant method for scalar equations.

Their calling sequences and the tuples they return are very similar to all the codes from this book. All our solvers return a tuple with the solution, the history of the iteration, flags for success or failure, and (optionally) the entire history of the solution.

We will continue this pattern for the entire book, discussing theory and algorithms at the beginning of each chapter together with a few examples. In the notebook section we will cover the software in more detail and go deeper into examples.

The solution history for scalar equations is small, and returning it is the default. In the later chapters on systems of equations, we do not return the solution history by default and discourage your asking for it. The solution history might take a lot of space to store and also, especially in Julia, have a severe penalty for allocations.

1.10.1 ▪ Newton's Method

nsolsc.jl is the scalar Newton solver. We will begin, as we will in all the software sections, by looking at the documentation (docstrings) in the code.

```
[2]: ?nsolsc
```

```
[2]: nsolsc(f,x0, fp=difffp; rtol=1.e-6, atol=1.e-12, maxit=10,
       solver="newton", sham=1, armmax=10, resdec=.1, dx=1.e-7,
       armfix=false, pdata=nothing, printerr=true, keepsolhist=true,
       stagnationok=false)

     C. T. Kelley, 2022

     Newton's method for scalar equations. Has most of the features
     a code for systems of equations needs. This is a wrapper for a
     call to nsol.jl, the real code for systems.

     Input:

     f: function

     x0: initial iterate

     fp: derivative. If your derivative function is fp, you give
     me its name. For example fp=foobar tells me that foobar is
     your function for the derivative. The default is a forward
     difference Jacobian that I provide.

     Keyword Arguments (kwargs):
```

rtol, atol: real and absolute error tolerances

maxit: upper bound on number of nonlinear iterations

solver:

Your choices are "newton" (default) or "chord". However, you
have sham at your disposal only if you choose newton. "chord"
will keep using the initial derivative until the iterate
converges, uses the iteration budget, or the line search fails.
It is not the same as sham=Inf, which is smarter.

If you use secant and your initial iterate is poor, you have
made a mistake. I will help you by driving the line search with
a finite difference derivative.

sham:

This is the Shamanskii method. If sham=1, you have Newton.
The iteration updates the derivative every sham iterations.
The convergence rate has local q-order sham+1 if you only
count iterations where you update the derivative. You need
not provide your own derivative function to use this option.
sham=Inf is chord only if chord is converging well.

armmax: upper bound on stepsize reductions in linesearch

resdec: target value for residual reduction.

The default value is .1. In the old MATLAB codes it was .5. I
only turn Shamanskii on if the residuals are decreasing rapidly,
at least a factor of resdec, and the line search is quiescent.
If you want to eliminate resdec from the method (you don't)
then set resdec = 1.0 and you will never hear from it again.

dx:

This is the increment for forward difference, default = 1.e-7.
dx should be roughly the square root of the noise in the
function.

armfix:

The default is a parabolic line search (ie false). Set to true
and the stepsize will be fixed at .5. Don't do this unless you
are doing experiments for research.

pdata:

precomputed data for the function/derivative. Things will go
better if you use this rather than hide the data in global
variables within the module for your function/derivative. If
you use this option your function and derivative must take pdata
as a second argument. eg f(x,pdata) and fp(x,pdata)

printerr:

I print a helpful message when the solver fails. To suppress
that message set printerr to false.

keepsolhist:

Set this to true to get the history of the iteration in the
output tuple. This is on by default for scalar equations and
off for systems. Only turn it on if you have use for the data,
which can get REALLY LARGE.

stagnationok:

Set this to true if you want to disable the line search and
either observe divergence or stagnation. This is only useful
for research or writing a book.

Output:

A named tuple (solution, functionval, history, stats, idid,
errcode, solhist) where

solution = converged result functionval = F(solution) history
= the vector of residual norms ($||F(x)||$) for the iteration
stats = named tuple of the history of (ifun, ijac, iarm), the
number of functions/derivatives/steplength reductions at each
iteration.

I do not count the function values for a finite-difference
derivative because they count toward a Jacobian evaluation. I
do count them for the secant method model.

idid=true if the iteration succeeded and false if not.

errcode = 0 if the iteration succeeded = -1 if the initial
iterate satisfies the termination criteria = 10 if no
convergence after maxit iterations = 1 if the line search failed

solhist:

This is the entire history of the iteration if you've set
keepsolhist=true

nsolsc builds solhist with a function from the Tools directory.
For systems, solhist is an N x K array where N is the length
of x and K is the number of iteration + 1. So, for scalar
equations (N=1), solhist is a row vector. Hence the use of
solhist' in the example below.

Examples for nsolsc.jl

```
julia> nsolout=nsolsc(atan,1.0;maxit=5,atol=1.e-12,rtol=1.e-12);
```

```
julia> nsolout.history
6-element Array{Float64,1}:
 7.85398e-01
 5.18669e-01
 1.16332e-01
 1.06102e-03
 7.96200e-10
 2.79173e-24
```

If you have an analytic derivative, I will use it.

```
julia> fs(x)=x^2-4.0; fsp(x)=2x;

julia> nsolout=nsolsc(fs,1.0,fsp; maxit=5,atol=1.e-9,rtol=1.e-9);

julia> [nsolout.solhist'.-2 nsolout.history]
6×2 Array{Float64,2}:
 -1.00000e+00   3.00000e+00
  5.00000e-01   2.25000e+00
  5.00000e-02   2.02500e-01
  6.09756e-04   2.43940e-03
  9.29223e-08   3.71689e-07
  2.22045e-15   8.88178e-15
```

You can also use anonymous functions

```
julia> nsolout=nsolsc(atan,10.0,x -> 1.0/(1.0+x^2);
atol=1.e-9,rtol=1.e-9);

julia> nsolout.history
8-element Vector{Float64}:
 1.47113e+00
 1.19982e+00
 1.10593e+00
 6.48297e-01
 2.56983e-01
 1.19361e-02
 1.13383e-06
 9.71970e-19
```

Input Let's begin with the calling sequence for the solver.

```
function nsolsc(
    f,
    x0,
    fp = difffp;
    rtol = 1.e-6,
    atol = 1.e-12,
```

```
      maxit = 10,
      solver = "newton",
      sham = 1,
      armmax = 5,
      resdec = 0.1,
      dx = 1.e-7,
      armfix = false,
      pdata = nothing,
      printerr = true,
      keepsolhist = true,
      stagnationok = false,
)
```

The arguments before the semicolon are required except for the function that evaluates the derivative. If you leave it out, **nsolsc** defaults to a forward difference approximation. We are solving $f(x) = 0$ and the solver needs f and the initial iterate x. The arguments after the semicolon are **keyword arguments**, usually referred to as **kwargs**, which is not a German cheese product. The semicolon is **very useful**. You may leave it out in recent versions of Julia, but it is good practice to leave it in to remind yourself about which arguments are optional and which are not. The good news about kwargs is that you may use any of them without worrying about the others, which will take their default values. So

```
nsolout0 = nsolsc(atan, 1.0)
```

```
nsolout1=nsolsc(atan, 3.0; sham=2, resdec=.5)
```

are all correct.

You have seen many of the kwargs before. The relative and absolute error tolerances, the solver, the parameters for the Shamanskii method and line search should be familiar. The new things are **resedec, dx, armfix, pdata, printerr**, and **keepsolshist**. For example, the derivative is updated every **sham** iteration. The default for Newton's method is sham=1 for scalar equations. We do something else for systems and will explain that in Chapter 2.

The default for derivative evaluation is a forward difference derivative with different increment $h = dx * |x| + 10^{-8}$. That is an internal function **difffp**. If you have an analytic derivative, say **fpanal.jl**, then set fp=fpanal and the solver will use your derivative.

We have mentioned the solution history before. Please leave **keepsolhist** alone unless there's a good reason to change it. It is set to true for scalar codes and false for the solvers in the following chapters.

We use the kwarg **resdec** to manage Shamanskii iterations. In this scalar code, it is used for some examples and to prepare you for its more serious use in the codes for systems of equations. In **nsolsc.jl** the default solver is Newton's method (so sham=1). **nsolsc.jl** with sham=2 is the Shamanskii method with a derivative update every two iterations. **But** we safeguard the skipping of the update by doing the update anyhow if (1) the line search fails on the first attempt (i.e., with step length = 1) or (2) the residual decrease is more than resdec. If you want to eliminate the second of these, set `resdec = 1`.

The only exception to the first criterion is the chord method. If you set `solver="chord"`, then you will get the chord method. That's in there for research and a few internal tests of the code. If you set sham=Inf, then you'll get the chord method with derivative updates when a step length of 1 fails to produce sufficient decrease or the reduction in residuals is not enough.

With the secant method, we compute the secant approximation to the derivative with every iteration. If the secant iterations are far apart, the approximation can be poor and the line search can fail. This is a risk you assume with all secant methods.

pdata is optional precomputed data for f. This becomes important in the subsequent chapters, where we will give many examples.

printerr = false turns off error messages.

Output   The output of all the solvers is a tuple. This is a data structure in Julia that can pack different structures (including more tuples) in a single data structure. It's a good way to manage complex output.

Here is a simple example of how to use the output tuple. I'll find the root $x^* = 0$ of $f(x) = \tan^{-1}(x)$ and look at the iteration statistics.

```
[3]:  tanout=nsolsc(atan,10.0);
      solution=tanout.solution;
      println("Solution = $solution.")

      functionval=tanout.functionval;
      println("Function = $functionval.")

      ifun=tanout.stats.ifun;
      println("Function call history = $ifun.")

      ijac=tanout.stats.ijac;
      println("Derivative call history = $ijac.")

      iarm=tanout.stats.iarm;
      println("Step size reduction history = $iarm.")

      # Now print the residual norm history
      tanout.history
```

```
Solution = -1.13371e-06.
Function = -1.13371e-06.
Function call history = [1, 3, 2, 2, 1, 1, 1].
Derivative call history = [0, 1, 1, 1, 1, 1, 1].
Step size reduction history = [0, 2, 1, 1, 0, 0, 0].
```

```
[3]:  7-element Vector{Float64}:
         1.47113e+00
         1.19982e+00
         1.10593e+00
         6.48294e-01
```

```
2.56979e-01
1.19356e-02
1.13371e-06
```

Note that I put a semicolon after calls to the solvers to stop printing the output after the call. This is exactly the same as the semicolon in MATLAB. I've taken the defaults for everything, so the iteration terminated when the residual was about 10^{-6}. Note that the solution history is part of the output. You can examine the history directly by asking for it. I store the solution history as a row vector to make the scalar equation code consistent with the one for systems, where the iterations are the columns of the solution history. Hence I print the transpose to make the results easier to look at.

```
[4]: tanout.solhist'
```

```
[4]: 7×1 adjoint(::Matrix{Float64}) with eltype Float64:
       1.00000e+01
       2.57080e+00
      -1.99394e+00
       7.57517e-01
      -2.62790e-01
       1.19362e-02
      -1.13371e-06
```

You can (and should) turn this off for systems of equations where the solution history can become a massive storage burden. It may be useful if you want to visualize the history with a movie, but is rarely worth the trouble. The solution and final function values are

```
[5]: [tanout.solution tanout.functionval]
```

```
[5]: 1×2 Matrix{Float64}:
      -1.13371e-06  -1.13371e-06
```

Looking at the iteration history can help you find errors or convince you that things are going well. For example, since $\mathrm{atan}'(0) = 1$, we would expect the solution and final function values to be the same, and they are. One simple thing we will do often is to plot a semilog graph of the residual history. If we are comparing methods, it is a good idea to plot $|f(x_n)|/|f(x_0)|$ so that all the plots will begin at the same point. Figure 1.8 is a simple version of that. You should always label axes, so I did. I make the curves black to save on publication costs for the print book.

```
[6]: residdata=tanout.history./tanout.history[1];
     sizehist=length(residdata)
     semilogy(0:sizehist-1,residdata,"k-");
     xlabel("iterations");
     ylabel("relative residual");
```

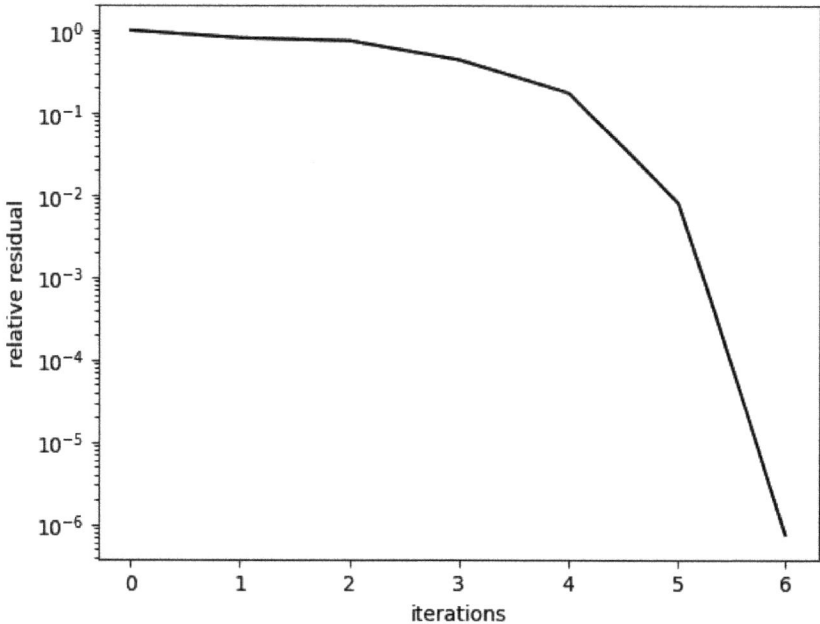

Figure 1.8: Residual history for atan.

The user interface for PyPlot is very similar to plotting in MATLAB, which is why I use it. The stats array has the data for function and derivative evaluations and the number of step size reductions in the line search. As you can see, the line search had some work to do in this example. If the iteration is not doing what you expect, a quick look at stats may help.

```
[7]: ifun=tanout.stats.ifun;
     println("Function call history = $ifun.")

     ijac=tanout.stats.ijac;
     println("Derivative call history = $ijac.")

     iarm=tanout.stats.iarm;
     println("Step size reduction history = $iarm.")
```

```
Function call history = [1, 3, 2, 2, 1, 1, 1].
Derivative call history = [0, 1, 1, 1, 1, 1, 1].
Step size reduction history = [0, 2, 1, 1, 0, 0, 0].
```

Here we see that the line search was active on iterations 1, 2, and 3 and every step size reduction costs a function evaluation like it's supposed to.

1.10.2 ▪ The Secant Method

We will now looks at the docstrings for **secant.jl**. The interface is very similar to **nsolsc.jl**. The only differences are that there is no reason to decide to reevaluate the derivative or not, since the secant method computes the model derivative with every iteration, so kwargs like **sham** are missing. The output tuple is essentially the same as the other codes from this book.

```
[8]: ?secant
```

```
[8]: secant(f,x0; rtol=1.e-6, atol=1.e-12, maxit=10, armmax=10,
     armfix=false, pdata=nothing, printerr=true, keepsolhist=true,
     stagnationok=false)

     C. T. Kelley, 2022

     The secant method for scalar equations.

     Input:

     f:  function

     x0:  initial iterate

     Keyword Arguments (kwargs):

     rtol, atol:  real and absolute error tolerances

     maxit:  upper bound on number of nonlinear iterations

     If you use secant and your initial iterate is poor, you have
     made a mistake.  You will get an error message.

     armmax:  upper bound on stepsize reductions in linesearch

     armfix:

     The default is a parabolic line search (ie false).  Set to true
     and the stepsize will be fixed at .5.  Don't do this unless you
     are doing experiments for research.

     printerr:

     I print a helpful message when the solver fails.  To suppress
     that message set printerr to false.

     keepsolhist:

     Set this to true to get the history of the iteration in the
     output tuple.  This is on by default for scalar equations and
     off for systems.  Only turn it on if you have use for the data,
     which can get REALLY LARGE.

     stagnationok:

     Set this to true if you want to disable the line search and
     either observe divergence or stagnation.  This is only useful
     for research or writing a book.

     Output:
```

A named tuple (solution, functionval, history, stats, idid, errcode, solhist) where

solution = converged result functionval = F(solution) history = the vector of residual norms (||F(x)||) for the iteration stats = named tuple of the history of (ifun, ijac, iarm), the number of functions/derivatives/steplength reductions at each iteration. For the secant method, ijac = 0.

idid=true if the iteration succeeded and false if not.

errcode = 0 if the iteration succeeded = -1 if the initial iterate satisfies the termination criteria = 10 if no convergence after maxit iterations = 1 if the line search failed

solhist:

This is the entire history of the iteration if you've set keepsolhist=true

secant builds solhist with a function from the Tools directory. For systems, solhist is an N x K array where N is the length of x and K is the number of iteration + 1. So, for scalar equations (N=1), solhist is a row vector. Hence the use of solhist' in the example below.

Example for secant.jl

```
julia> secout=secant(atan,1.0;maxit=6,atol=1.e-12,rtol=1.e-12);


julia> secout.history
7-element Array{Float64,1}:
 7.85398e-01
 5.18729e-01
 5.39030e-02
 4.86125e-03
 4.28860e-06
 3.37529e-11
 2.06924e-22
```

Now we can play with the options for **nsolsc.jl** and **secant.jl**. Let's tighten the tolerances and compare three solvers. Newton's method and the secant method successfully found the solution. The chord method failed to converge and gave an error message.

```
[9]:  tannewt=nsolsc(atan,1.0; rtol=1.e-12, atol=1.e-12);
      tansec=secant(atan,1.0; rtol=1.e-12, atol=1.e-12);
      tanchord=nsolsc(atan,1.0; rtol=1.e-12, atol=1.e-12,
          solver="chord");
```

```
Maximum iterations (maxit) of 10 exceeded
Convergence failure: residual norm too large  2.46512e-01
```

```
Try increasing maxit and checking your function and
                    Jacobian for bugs.
Give the history array a look to see what's happening.
```

Let's examine the history array for the chord iteration.

[10]: `tanchord.history`

[10]: 11-element Vector{Float64}:
       7.85398e-01
       5.18669e-01
       4.36525e-01
       3.86104e-01
       3.50593e-01
       3.23686e-01
       3.02326e-01
       2.84806e-01
       2.70083e-01
       2.57474e-01
       2.46512e-01

Yes, that's going nowhere at all. The problem is that the derivative at $x_0 = 1$ is not good enough to make the chord iteration converge well. It would get there eventually, but only after far more iterations than you or I have the patience for. We'll plot the results in Figure 1.9 and see what happened.

[11]:
```
residdatan=tannewt.history./tannewt.history[1];
sizehistn=length(residdatan);

residdatas=tansec.history./tansec.history[1];
sizehists=length(residdatas);

residdatac=tanchord.history./tanchord.history[1];
sizehistc=length(residdatac);

semilogy(0:sizehistn-1,residdatan,"k-",
0:sizehists-1,residdatas,"k--",
0:sizehistc-1,residdatac,"k-."
);
legend(["Newton","Secant","Chord"]);
xlabel("iterations");
ylabel("relative residual");
```

This is exactly what you'd expect. The histories for the secant and Newton iterations have the downward concavity of superlinear convergence. Newton's method takes fewer non-linear iterations (but only by one) while the secant method uses fewer calls to functions. Remember that Newton does a function and derivative evaluation at each step, while the secant iteration only needs a single function evaluation per step.

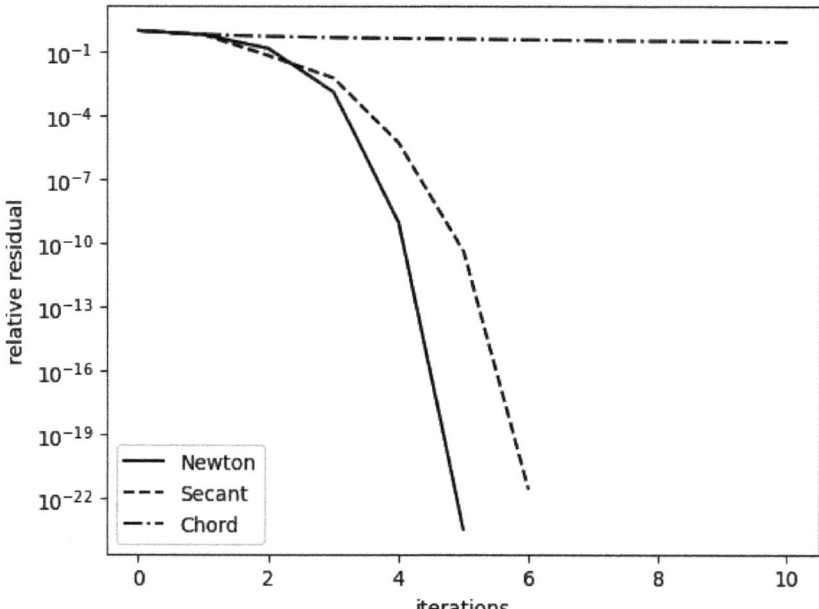

Figure 1.9: Global convergence history for atan.

We've seen how to run the solver, pick methods, and look at the output. We are almost done with this section and ready to move on to systems of equations in Chapter 2. The last thing to do is talk about how to package your function. Suppose you have Julia code for your function (**myfun.jl**) and its derivative (**myfunp.jl**). Once in the REPL or a notebook, you can put these functions where Julia can find them with the **include** command. You'd do

```
include("myfun.jl")
include("myfunp.jl")
```

and then could invoke the solvers. This is far from ideal for many reasons.

Julia is a compiled language. The code for your functions is compiled when you invoke the **include** command. If you make a change to **myfun.jl**, you would have to include it again before you could use it.

Another way is to use a module to contain all the code you need for your function. **My-Function.jl** is an example of such a module. If your function has many subfunctions and only a few of them are used outside of the function itself, putting them in a module is a very good idea. That is why all of the examples in this book are in the **TestProblems** submodule of SIAMFANLEquations.

```
module MyFunction

export myfun
export myfunp

function myfun(x)
```

```
        fun=cos(x)-x
        return fun
    end

    function myfunp(x)
        funp=-sin(x)-1
        return funp
    end
end
```

If you use modules and the **Revise.jl** [92] package, then changes in your module will be immediately recompiled and usable in the REPL. using Revise belongs in the **startup.jl** file of every Julia programmer. Put it on the first line of **startup.jl** to get it to work properly.

I've put **MyFunction.jl** in the **src/Chapter1** directory of NotebookSIAMFANL. You will be able to use it if you've run the include("fanote_init.jl") command in the first code cell. However, if it's your own module, you will need to make sure that it is in your **LOAD_PATH** and then type a using command in the REPL. In this case that command would be

```
[12]: using NotebookSIAMFANL.MyFunction
```

To see if the functions are where they need to be, we'll evaluate them.

```
[13]: x=pi;
      y=myfun(x);
      yp=myfunp(x);
      [y,yp]
```

```
[13]: 2-element Vector{Float64}:
       -4.14159e+00
       -1.00000e+00
```

We now have both **myfun** and its derivative **myfunp** ready to go. We will compare Newton's method with an analytic derivative to the iteration with a forward difference derivative (the default).

```
[14]: # Solve the equation twice.
      # Once with a finite difference derivative ...
          myfunforwarddiff=nsolsc(myfun,1.0);
      # and again with an analytic derivative.
          myfunanalytic=nsolsc(myfun,1.0,myfunp);
      # Now subtract the residual histories.
          myfunforwarddiff.history-myfunanalytic.history
```

```
[14]: 4-element Vector{Float64}:
       0.00000e+00
       6.53120e-09
       4.36679e-10
       1.00131e-12
```

There's no significant difference. We will return to this in the subsequent chapters.

Finally we will see how we circumvented the normal termination criteria to generate **Figure 1.3**. The trick is to use the **stagnationok** keyword. The call to Newton's method looked like

```
nnout=nsolsc(ftanx,4.5; maxit=14, rtol=1.e-17, atol=1.e-17,
        printerr=false, stagnationok=true)
```

Setting `stagnationok = true` turned off termination when the line search failed to obtain sufficient decrease, which would give more sensible results. The purpose of this option is to do experiments, and it's best left set to false (the default). A better version of **Figure 1.3** would do this, but not show stagnation as vividly. **fig1dot3b.jl** generates Figure 1.10 and does that.

[15]: 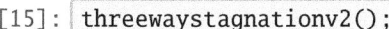 `threewaystagnationv2();`

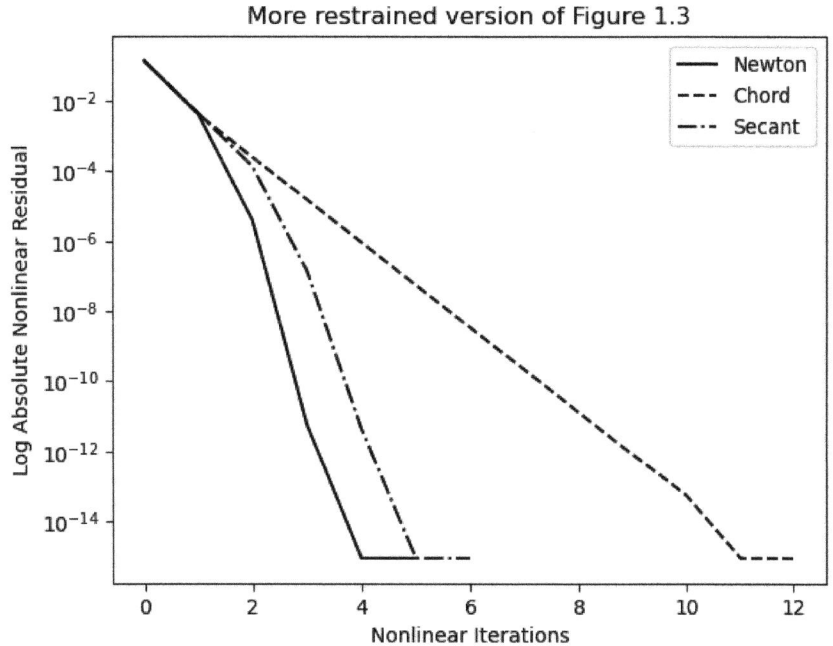

Figure 1.10: Stagnation.

1.10.3 ▪ Pseudo-transient Continuation

ptcsolsc.jl is the scalar pseudo-transient continuation solver. As usual, we will start with the docstrings.

[16]: `?ptcsolsc`

[16]: `ptcsolsc(f, x0, fp=difffp; rtol=1.e-6, atol=1.e-12, maxit=100,`

delta0=1.e-6, dx=1.e-7, pdata=nothing, printerr = true,
keepsolhist=true)

C. T. Kelley, 2022

Scalar pseudo-transient continuation solver. PTC is designed to
find stable steady state solutions of

dx/dt = - f(x)

The scalar code is a simple wrapper around a call to ptcsol.jl,
the PTC solver for systems.

-> PTC is ABSOLUTELY NOT a general purpose nonlinear solver.

Input:

f: function

x: initial iterate/data

fp: derivative. If your derivative function is fp, you give
me its name. For example fp=foobar tells me that foobar is
your function for the derivative. The default is a forward
difference Jacobian that I provide.

Keyword Arguments:

rtol, atol: real and absolute error tolerances

maxit: upper bound on number of nonlinear iterations. This
is coupled to delta0. If your choice of delta0 is too small
(conservative) then you'll need many iterations to converge and
will need a larger value of maxit.

delta0: initial pseudo time step. The default value of 1.e-3
is a bit conservative and is one option you really should play
with. Look at the example where I set it to 1.0!

dx: default = 1.e-7

difference increment in finite-difference derivatives
h=dx*norm(x)+1.e-6

pdata:

precomputed data for the function/derivative. Things will go
better if you use this rather than hide the data in global
variables within the module for your function/derivative. If
you use this option your function and derivative must take pdata
as a second argument. eg f(x,pdata) and fp(x,pdata)

printerr: default = true

I print a helpful message when the solver fails. To suppress
that message set printerr to false.

keepsolhist: if true you get the history of the iteration in the output tuple. This is on by default for scalar equations and off for systems. Only turn it on if you have use for the data, which can get REALLY LARGE.

Output: A tuple (solution, functionval, history, idid, errcode, solhist) where history is the array of absolute function values |f(x)| of residual norms and time steps. Unless something has gone badly wrong, delta approx |f(x_0)|/|f(x)|.

idid=true if the iteration succeeded and false if not.

errcode = 0 if the iteration succeeded = -1 if the initial iterate satisfies the termination criteria = 10 if no convergence after maxit iterations

solhist=entire history of the iteration if keepsolhist=true

ptcsolsc builds solhist with a function from the Tools directory. For systems, solhist is an N x K array where N is the length of x and K is the number of iteration + 1. So, for scalar equations (N=1), solhist is a row vector. Hence I use [ptcout.solhist' ptcout.history] in the example below.

If the iteration fails it's time to play with the tolerances, delta0, and maxit. You are certain to fail if there is no stable solution to the equation.

Examples for ptcsolsc

```
julia> ptcout=ptcsolsc(sptest,.2;delta0=2.0,rtol=1.e-3,atol=1.e-3);

julia> [ptcout.solhist' ptcout.history]
7x2 Array{Float64,2}:
 2.00000e-01  9.20000e-02
 9.66666e-01  4.19962e-01
 8.75086e-01  2.32577e-01
 7.99114e-01  1.10743e-01
 7.44225e-01  4.00926e-02
 7.15163e-01  8.19395e-03
 7.07568e-01  4.61523e-04
```

The example $sptest(x) = x^3 - 0.5x$ is in the TestProblems submodule in the file **Test-Problems/Scalars/spitchfork.jl**.

That file contains four functions. The ones of interest here are $sptest$ and its derivative $sptestp$. We will illustrate how **ptcsolsc** works by looking at the effects of larger values of δ_0. Using $\delta_0 = \infty$ is Newton's method and you'll get convergence to $x^* = 0$, the unstable solution, if you do that. We will test a few values of δ_0 and plot the solution histories. As you can see, even a large value of δ_0 will keep the iteration away from the unstable branch, but convergence to the stable branch will take a very long time. The iteration fails for two of the runs because the maximum iteration count limit was hit. The iterations for $\delta_0 = 1$

and $\delta_0 = 10$ do converge to the stable solution within the 200 iterations I gave the solver. You are welcome to increase the iteration limit and see how painful it can be.

We have a utility **plothist.jl** in src/Tools in the Notebook repository. That utility plots histories like the one in Figure 1.11. We use **plothist.jl** to plot the solution x in this case so we can observe the convergence to $x^* = \sqrt{1/2}$.

[17]:
```julia
## Set up the problem
x0=.1
ustable=.5*sqrt(2.0)
uunstable=0.0
## for a few curated values of delta0
dtlist=[1.0, 1.e1, 1.e2, 1.e4]
## and collect the results.
outdata=[]
for id=1:4
    dti=10.0^(1-id)
    ptcdata1=ptcsolsc(sptest,x0;
        delta0=dti, rtol=1.e-6, maxit=200)
    push!(outdata,ptcdata1.solhist')
end
## Get some labels for the plots.
labels=[
L"$\delta_0 = 1.0$",
L"$\delta_0 = 10$",
L"$\delta_0 = 10^{2}$",
L"$\delta_0 = 10^{4}$"
]
## Make the figure.
plothist(outdata,labels,"solution");
```

```
PTC failure; increase maxit and/or delta0
Residual norm =  9.13545e-02
Current values: maxit  =  200,  delta0 = 1.00000e-02
Give the history array a look to see what's happening.

PTC failure; increase maxit and/or delta0
Residual norm =  5.35783e-02
Current values: maxit  =  200,  delta0 = 1.00000e-03
Give the history array a look to see what's happening.
```

The message so far is that if δ_0 is either too small or too large, convergence will suffer and that too small is better than too large. One reason for this is the theoretical result that you will converge to the steady-state solution you want, namely, $\mathbf{x}^* = \lim_{t \to \infty} \mathbf{x}(t)$ if δ_0 is sufficiently small.

If δ_0 is too large, the results can be inconsistent. In the example above, the two convergent iterations found the correct steady-state solution $x^* = \sqrt{1/2}$. However,

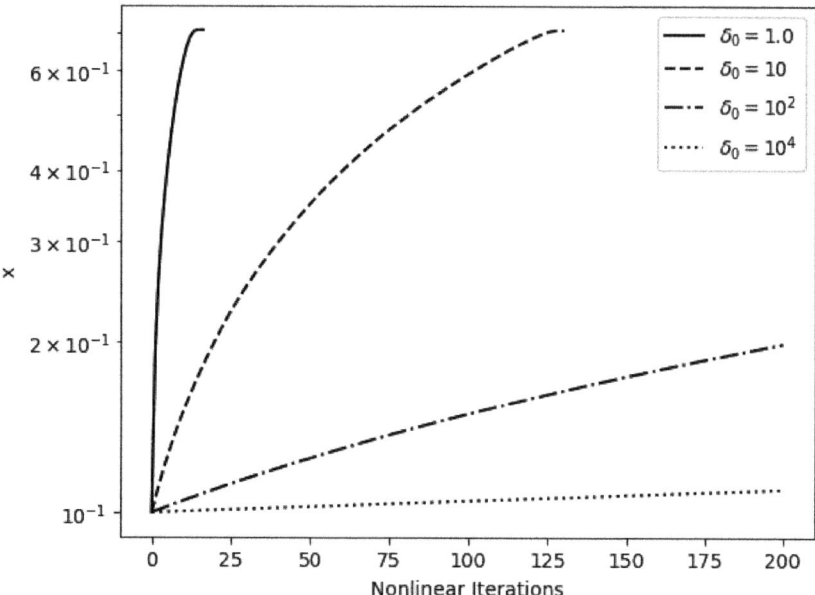

Figure 1.11: ΨTC iteration history.

```
[18]: ptcoutwrong=ptcsolsc(sptest,x0; delta0=3.0);
      [ptcoutwrong.solhist' ptcoutwrong.history]
```

```
[18]: 8×2 Matrix{Float64}:
          1.00000e-01   4.90000e-02
         -2.58537e-01   1.11987e-01
         -5.00754e-01   1.24811e-01
         -6.14083e-01   7.54723e-02
         -6.80014e-01   2.55555e-02
         -7.04098e-01   2.98965e-03
         -7.07065e-01   4.17213e-05
         -7.07107e-01   8.13958e-09
```

That's something to think about, isn't it? Note that the residual norm **increased** on its way to the wrong steady state (wrong means that it is not $\lim_{t\to\infty} x(t)$).

There's one more thing to say about the sptest example. It is a special case ($\lambda = 1/2$) of

$$f(x, \lambda) = x^3 - \lambda x.$$

If we plot **all** the solutions of $f(x, \lambda) = 0$ we get a **bifurcation diagram**. The solution sets are the line $x = 0$ and the parabola $\lambda = x^2$. Remember that $x = 0$ is stable for $\lambda < 0$ and unstable for $\lambda > 0$. We will label the stable and unstable branches in Figure 1.12.

```
[19]: PitchFork1();
```

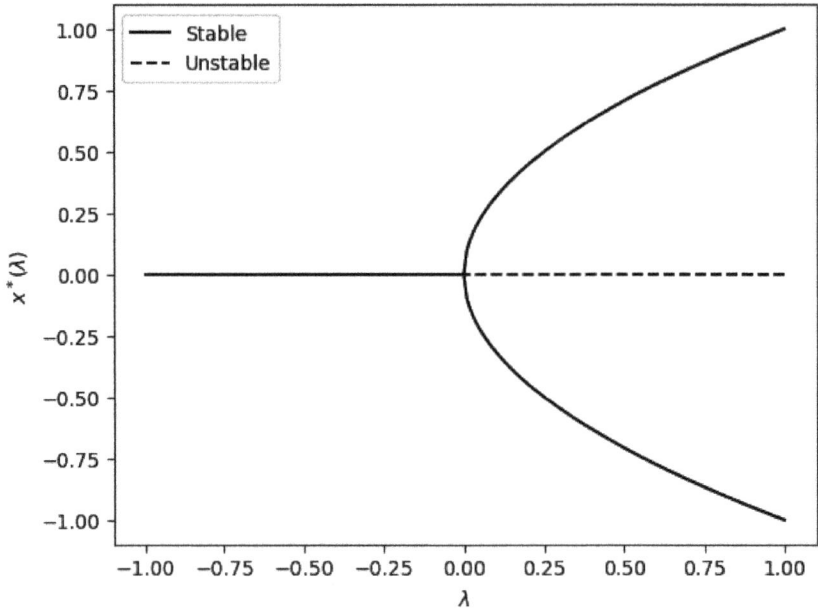

Figure 1.12: Pitchfork bifurcation.

Moving from the left you see that at $\lambda = 0$ the stable solution **bifurcates** into two stable branches. The implicit function theorem [106] says that this can only happen if $\partial f(0,0)/\partial x = 0$. This kind of singularity is called, for obvious reasons, a **pitchfork bifurcation** [76, 106]. We will see a more interesting example in the subsequent chapters.

1.11 ▪ Projects

1.11.1 ▪ Estimating the q-Order

One can examine the data in the history vector to estimate the q-order in the following way. If $\mathbf{x}_n \to \mathbf{x}^*$ with q-order p, then one might hope that

$$\|\mathbf{F}(\mathbf{x}_{n+1})\| \approx K\|\mathbf{F}(\mathbf{x}_n)\|^p$$

for some $K > 0$. If that happens, then, as $n \to \infty$,

$$\log(\|\mathbf{F}(\mathbf{x}_{n+1})\|) \approx p\log(\|\mathbf{F}(\mathbf{x}_n)\|)$$

and so

$$p \approx \frac{\log(\|\mathbf{F}(\mathbf{x}_{n+1})\|)}{\log(\|\mathbf{F}(\mathbf{x}_n)\|)}.$$

Hence, by looking at the history array, we can estimate p.

Here are a few examples for Newton's method. We'll look at $f(x) = \cos(x) - x$ from the MyFunction module and $f(x) = \operatorname{atan}(x)$ and examine the logs of the residuals.

```
[20]: myfunout=nsolsc(myfun,0.1)
      loghist=log.(myfunout.history)
      nn=length(loghist)
      qest=loghist[2:nn]./loghist[1:nn-1]
```

```
[20]: 4-element Vector{Float64}:
          1.07643e+01
          3.91318e+00
          2.43526e+00
          2.17677e+00
```

This is pretty good. Since the iteration should be quadratically convergent, we would expect the sequence to converge to 2. We are not aiming for high precision results here and should be happy.

And now for $f(x) = \text{atan}(x)$:

```
[21]: tanout=nsolsc(atan,1.0)
      loghist=log.(tanout.history)
      nn=length(loghist)
      qest=loghist[2:nn]./loghist[1:nn-1]
```

```
[21]: 4-element Vector{Float64}:
          2.71765e+00
          3.27699e+00
          3.18343e+00
          3.05923e+00
```

Wait a minute! What happened to the q-order of 2? In fact, the q-order for this problem is 3! Can you figure out why?

Your job is to

  • apply this idea for the Newton, Shamanskii, secant, and chord methods for the example problems in this chapter;
  • try it for $\sin(x) = 0$ with an initial iterate of $x_0 = 3$.

Are the estimated q-orders consistent with the theory? Can you explain the q-orders that you observe?

1.11.2 ▪ Singular Problems

Solve $f(x) = x^2 = 0$ with Newton's method, the chord method, and the secant method. Use $x_0 = 1$ for the initial iterate. Try the alternative iteration

$$x_{n+1} = x_n - 2f'(x_n)^{-1}f(x_n).$$

Look at the history vectors from the two cases. Can you explain your observations?

1.11.3 ▪ Non-Lipschitz Derivative

Solve $f(x) = 0$ where $f(x) = x + |x|^{3/2}$ with Newton's method. What is a reasonable initial iterate? Do you see q-superlinear convergence? Explain your results and then look at [105] to see if you got it right.

1.11.4 ▪ A Test Drive

A numerical analyst buys a German sports car for $50,000. He puts $10,000 down and takes a 7-year installment loan to pay the balance. If the monthly payments are $713.40, what is the interest rate? Assume monthly compounding.

Chapter 2

Finding the Newton Step with Gaussian Elimination

Files for This Chapter

- From the Package repository:
    - Solvers using direct methods: **/src/Solvers**
        * Newton's method: **nsol.jl**
        * Pseudo-transient continuation: **ptcsol.jl**
    - Test Problems: **/src/TestProblems/Systems**
        * Two-dimensional problem: **simple!.jl**
        * H-equation: **Hequation.jl**
        * Convection-diffusion equation: **EllipticPDE.jl** and **PDE_Tools.jl**
        * Buckling beam: **FBeam!.jl**
        * Two-point boundary value problem: **Fbvp!.jl**
    - Examples
        * Initial value problem for the buckling beam: **ivpBeam.jl**
        * Pseudo-transient continuation for the buckling beam: **ptcBeam.jl**
- From the Notebook repository: **/src/Chapter2**
  Julia codes that generate the figures and tables

2.1 ▪ Direct Methods for Solving Linear Equations

Direct methods for solving the equation for the Newton step are a good idea if

- the Jacobian can be computed and **stored** efficiently, and
- the cost of the factorization of the Jacobian is not excessive, or

49

- iterative methods do not converge for your problem.

Even when direct methods work well, Jacobian factorization and storage of that factorization may be more expensive than a solution by iteration. However, direct methods are more robust than iterative methods and do not require your worrying about the possible convergence failure of an iterative method or preconditioning.

If the linear equation for the Newton step is solved exactly and the Jacobian is computed and factored with each nonlinear iteration (i.e., $\eta = 0$ in Algorithm **nsolg**), one should expect to see q-quadratic convergence until finite precision effects produce stagnation (as predicted in Theorem 1.2). One can, of course, approximate the Jacobian or evaluate it only a few times during the nonlinear iteration, exchanging an increase in the number of nonlinear iterations for a dramatic reduction in the cost of the computation of the steps.

In many cases, as we pointed out in Chapter 1, one can store and factor $\mathbf{F}'$ in reduced precision with no loss in the quality of the solution. If, as we sometimes do in this book, one computes $\mathbf{F}$ in double precision (Float64), storing and factoring $\mathbf{F}'$ in single precision (Float32) cuts the linear algebra cost in half [113]. Our solvers enable you to do this with ease by allocating the storage for $\mathbf{F}'$ in single precision. We have a detailed example of this in the notebook in section 2.8.2.

In this chapter we solve the equation for the Newton step with Gaussian elimination. As is standard in numerical linear algebra (see [53, 75, 185, 195], for example), we distinguish between the factorization and the solve. The typical implementation of Gaussian elimination, called an **LU factorization**, factors the coefficient matrix $\mathbf{A}$ into a product of a permutation matrix and lower and upper triangular factors:

$$\mathbf{A} = \mathbf{PLU}.$$

The factorization may be simpler and less costly if the matrix has an advantageous structure (sparsity, symmetry, positivity, etc.) [7, 53, 57, 75, 185, 195].

The permutation matrix reflects row interchanges that are done during the factorization to improve stability. In Julia the best way to do this for dense matrices is to factor $\mathbf{A}$ **in-place**. This means that $\mathbf{L}$ and $\mathbf{U}$ are stored by **overwriting** $\mathbf{A}$ and the permutation is recorded in an integer vector. Julia's convention for a function that overwrites its input is to put a "!" after the name of the function. So you invoke the LU factorization with

```
B = lu!(A)
```

There is a subtle point here that really matters for Newton's method. While $\mathbf{B}$ and $\mathbf{A}$ share storage, they are different data structures. One can, for example, update $\mathbf{A}$ as a matrix, as one would do in a nonlinear solver if $\mathbf{A}$ is an approximation of the Jacobian. One cannot do that if one makes the call as

```
A = lu!(A)
```

because then you have changed the data type of $\mathbf{A}$.

For example, let

$$\mathbf{A} = \begin{pmatrix} 4 & 6 & 6 \\ 2 & 2 & 3 \\ 7 & 8 & 10 \end{pmatrix}.$$

The LU factorization returned by the Julia command **lu!** is

```
julia> B=lu!(A)
LU{Float64,Array{Float64,2}}
L factor:
3×3 Array{Float64,2}:
  1.00000e+00   0.00000e+00   0.00000e+00
  5.71429e-01   1.00000e+00   0.00000e+00
  2.85714e-01  -2.00000e-01   1.00000e+00
U factor:
3×3 Array{Float64,2}:
  7.00000e+00   8.00000e+00   1.00000e+01
  0.00000e+00   1.42857e+00   2.85714e-01
  0.00000e+00   0.00000e+00   2.00000e-01
```

The permutation is elsewhere in B. To see this

```
julia> B.L * B.U
3×3 Array{Float64,2}:
  7.00000e+00   8.00000e+00   1.00000e+01
  4.00000e+00   6.00000e+00   6.00000e+00
  2.00000e+00   2.00000e+00   3.00000e+00
```

indicating that $\mathbf{LU} = \mathbf{P}^T\mathbf{A}$. The permutation vector is

```
julia> B.p
3-element Array{Int64,1}:
 3
 1
 2
```

We use the output of **lu!** with the standard backslash notation. So to solve $\mathbf{Ax} = \mathbf{b}$ one would do this:

```
julia> B=lu!(A);
julia> s=B\b;
```

In the REPL, the semicolon suppresses output, just as it does in MATLAB.

We will ignore the permutation for the remainder of this chapter, but the reader should remember that it is important. Most linear algebra software [7,57] manages the permutation for you in some way.

The cost of an LU factorization of an $N \times N$ matrix is $N^3/3 + O(N^2)$ flops, where, following [57], we define a flop as an add, a multiply, and some address computations. The factorization is the most expensive part of the solution.

Following the factorization, one can solve the linear system as we described above. Internally that means to solve $\mathbf{As} = \mathbf{b}$ by solving the two triangular systems $\mathbf{Lz} = \mathbf{b}$ and $\mathbf{Us} = \mathbf{z}$. The cost of the two triangular solves is $N^2 + O(N)$ flops.

One final point, which we will return to soon, is that using **lu!** rather than **lu** is very important. The **lu** function performs the same factorization, but allocates new storage for the results rather than overwriting the original matrix. The penalty for allocating memory in Julia is severe and one must take care to avoid it.

2.2 ▪ The Newton–Armijo Iteration

Algorithm **newton** is an implementation of Newton's method that uses Gaussian elimination to compute the Newton step. The significant contributors to the computational cost are the computation and LU factorization of the Jacobian. The factorization can fail if, for example, $\mathbf{F}'$ is singular or highly ill-conditioned.

ALGORITHM 2.1.
newton$(\mathbf{F}, \mathbf{x}, \tau_a, \tau_r)$
    Evaluate $\mathbf{F}(\mathbf{x})$; $\tau \leftarrow \tau_r \|\mathbf{F}(\mathbf{x})\| + \tau_a$.
    while $\|\mathbf{F}(\mathbf{x})\| > \tau$ **do**
       Compute $\mathbf{F}'(\mathbf{x})$; factor $\mathbf{F}'(\mathbf{x}) = \mathbf{LU}$.
       if the factorization fails **then**
          report an error and terminate
       else
          solve $\mathbf{LUd} = -\mathbf{F}(\mathbf{x})$
       end if
       Find a step length λ using a polynomial model.
       $\mathbf{x} \leftarrow \mathbf{x} + \lambda \mathbf{d}$
       Evaluate $\mathbf{F}(\mathbf{x})$.
    end while

The Newton-like solvers in this chapter all follow this paradigm. The bulk of the work and the difference in the solvers are in the evaluation of the (approximate) Jacobian and the linear solve. If one uses direct methods, one can think of the evaluation and factor

Compute $\mathbf{F}'(\mathbf{x})$; factor $\mathbf{F}'(\mathbf{x}) = \mathbf{LU}$.

as a single operation that prepares a linear operator to compute the step.

2.3 ▪ Computing a Finite Difference Jacobian

The effort in the computation of the Jacobian can be substantial. In some cases one can compute the function and the Jacobian at the same time, and the Jacobian costs little more (see the example in section 2.7.2; also see section 2.5.2) than the evaluation of the function. However, if only function evaluations are available, then approximating the Jacobian by differences is one option (but see the project on automatic differentiation at the end of this chapter). As we said in Chapter 1, this usually causes no problems in the nonlinear iteration, and a forward difference approximation is almost always sufficient. One computes the forward difference approximation $(\nabla_h F)(x)$ to the Jacobian by columns. The jth column is

$$(\nabla_h \mathbf{F})(\mathbf{x})_j = \begin{cases} \dfrac{\mathbf{F}(\mathbf{x} + h\mathbf{e}_j) - \mathbf{F}(\mathbf{x})}{h}, & \mathbf{x} \neq 0, \\[2ex] \dfrac{\mathbf{F}(h\mathbf{e}_j) - \mathbf{F}(\mathbf{x})}{h}, & \mathbf{x} = 0. \end{cases} \qquad (2.1)$$

In (2.1) $\mathbf{e}_j$ is the unit vector in the jth coordinate direction. The difference increment h should be no smaller than the square root of the inaccuracy in $\mathbf{F}$. Each column of $\nabla_h \mathbf{F}$ requires one new function evaluation and, therefore, a finite difference Jacobian costs N function evaluations.

The difference increment in (2.1) should be **scaled**. Rather than simply perturb $\mathbf{x}$ by a difference increment h, roughly the square root of the error in $\mathbf{F}$, we multiply the perturbation by $\|\mathbf{x}\|$ [107]. Hence, if the error in $\mathbf{F}$ is ϵ, we use a difference increment of $h = \|\mathbf{x}\| dx$, where $dx \approx \sqrt{\epsilon}$, and $h = dx$ if $\mathbf{x} = 0$. You can adjust this in our solvers. The default in the solvers for this book is $dx = 10^{-7}$. We give an example of a problem where adjusting dx is useful in section 5.1.

2.4 ▪ The Chord and Shamanskii Methods

If the computational cost of a forward difference Jacobian is high ($\mathbf{F}$ is expensive and/or N is large) and if an analytic Jacobian is not available, it is wise to amortize this cost over several nonlinear iterations. The **chord method** from section 1.3 does exactly that. Recall that the chord method differs from Newton's method in that the evaluation and factorization of the Jacobian are done only once for $\mathbf{F}'(\mathbf{x}_0)$. The advantages of the chord method increase as N increases, since both the N function evaluations and the $O(N^3)$ work (in the dense matrix case) in the matrix factorization are done only once. So, while the convergence is q-linear and more nonlinear iterations will be needed than for Newton's method, the overall cost of the solve will usually be much less. The chord method is the solver of choice in many codes for stiff initial value problems [10, 20, 162], where the Jacobian may not be updated for several time steps.

Algorithms **chord** and **shamanskii** are special cases of **nsolg**. Global convergence problems have been ignored, so the step and the direction are the same, and the computation of the step is based on an LU factorization of $\mathbf{F}'(\mathbf{x})$ at an iterate that is generally not the current one.

ALGORITHM 2.2.
chord$(\mathbf{F}, \mathbf{x}, \tau_a, \tau_r)$
    Evaluate $\mathbf{F}(\mathbf{x})$; $\tau \leftarrow \tau_r \|\mathbf{F}(\mathbf{x})\| + \tau_a$.
    Compute $\mathbf{F}'(\mathbf{x})$; factor $\mathbf{F}'(\mathbf{x}) = \mathbf{LU}$.
    if the factorization fails **then**
        report an error and terminate
    else
        while $\|\mathbf{F}(\mathbf{x})\| > \tau$ **do**
            Solve $\mathbf{LUs} = -\mathbf{F}(\mathbf{x})$.
            $\mathbf{x} \leftarrow \mathbf{x} + \mathbf{s}$
            Evaluate $\mathbf{F}(\mathbf{x})$.
        end while
    end if

A middle ground is the **Shamanskii method** (see section 1.3) [175]. Here the Jacobian factorization and matrix function evaluation are done after every m computation of the step.

ALGORITHM 2.3.
shamanskii$(\mathbf{F}, \mathbf{x}, \tau_a, \tau_r, m)$

     Evaluate $\mathbf{F}(\mathbf{x})$; $\tau \leftarrow \tau_r \|\mathbf{F}(\mathbf{x})\| + \tau_a$.
     while $\|\mathbf{F}(\mathbf{x})\| > \tau$ **do**
         Compute $\mathbf{F}'(\mathbf{x})$; factor $\mathbf{F}'(\mathbf{x}) = \mathbf{LU}$.
         if the factorization fails **then**
             report an error and terminate
         end if
         for $p = 1 : m$ **do**
             Solve $\mathbf{LUs} = -\mathbf{F}(\mathbf{x})$.
             $\mathbf{x} \leftarrow \mathbf{x} + \mathbf{s}$
             Evaluate $\mathbf{F}(\mathbf{x})$; if $\|\mathbf{F}(\mathbf{x})\| \le \tau$ terminate.
         end for
     end while

If one counts as a complete iteration the full m steps between Jacobian computations and factorizations, the Shamanskii method converges q-superlinearly with **q-order** $m + 1$; i.e.,

$$\|\mathbf{x}_{n+1} - \mathbf{x}^*\| \le K \|\mathbf{x}_n - \mathbf{x}^*\|^{m+1}$$

for some $K > 0$. Newton's method, of course, is the $m = 1$ case. The default in **nsol.jl** is $m = 5$ for systems of equations. This is different from that in **nsolsc.jl**, where we use $m = 1$ for scalar equations. We will look into this and other ways to improve the performance of Newton's method in the notebook.

2.5 ▪ What Can Go Wrong?

The list in section 1.9 is complete, but it's worth thinking about a few specific problems that can arise when you compute the Newton step with a direct method. The major point to remember is that if you use an approximation to the Jacobian, then the line search can fail. You should think of the chord and Shamanskii methods as local algorithms, to which a code will switch after a Newton–Armijo iteration has resolved any global convergence problems. With the exception of the chord method, where we leave you on your own, our solvers update the Jacobian if the line search reduces the step length on the previous iteration.

2.5.1 ▪ Poor Jacobians

The chord method and other methods that amortize factorizations over many nonlinear iterations perform well because factorizations are done infrequently. This means that the Jacobians will be inaccurate, but if the initial iterate is good, the Jacobians will be accurate enough for the overall performance to be far better than a Newton iteration. However, if your initial iterate is far from a solution, this inaccuracy can cause a **line search to fail**. Even if the initial iterate is acceptable, the convergence may be slower than you'd like.

Our code **nsol.jl** (see section 2.8.1) watches for these problems and updates the Jacobian if either the line search is activated or the rate of reduction in the nonlinear residual is too slow.

2.5.2 ▪ Finite Difference Jacobian Error

The choice of finite difference increment h deserves some thought. You were warned in sections 1.9.3 and 1.9.4 that the difference increment in a forward difference approximation to a Jacobian or a Jacobian-vector product should be a bit more than the square root of the error in the function. Most codes, including ours, assume that the error in the function is on the order of floating point roundoff. If that assumption is not valid for your problem, the difference increment must be adjusted to reflect that. Check that you have scaled the difference increment to reflect the size of $\mathbf{x}$, as we did in (2.1). If the components of $\mathbf{x}$ differ dramatically in size, consider a change of independent variables to rescale them.

One might think that centered differences would be better, but that would be wrong [143]. The cost of a centered difference Jacobian is very high and there is no benefit for the nonlinear iteration. Another approach [132, 184] uses complex arithmetic to get higher-order accuracy. If $\mathbf{F}$ is smooth and can be evaluated for complex arguments, then you can get a second-order accurate directional derivative with a single function evaluation by using the formula

$$Im(\mathbf{F}(\mathbf{x} + ih\mathbf{u}))/h = \mathbf{F}'(\mathbf{x})\mathbf{u} + O(h^2). \tag{2.2}$$

One should use (2.2) with some care if there are errors in $\mathbf{F}$ and, of course, one should scale h.

One other approach to more accurate derivatives is automatic differentiation [80]. Automatic differentiation software takes as its input a code for $\mathbf{F}$ and produces a code for $\mathbf{F}$ and $\mathbf{F}'$. The derivatives are exact, but the codes are usually less efficient and larger than a hand-coded Jacobian program would be. Automatic differentiation software for C, C++, and FORTRAN is available from the US Department of Energy laboratories [86,94]. There are also many packages in Julia, for example, **Zygote.jl** [98], **ForwardDiff.jl** [164], and **ReverseDiff.jl** [163]. Automatic differentiation is far beyond the scope of this book, but we encourage the reader to investigate it and we have a project about that in section 2.9.2.

2.5.3 ▪ Poor Choice of δ_0 in ΨTC

We repeat the caution from sections 2.7.6 and 1.9.5 on the choice of δ_0 in ΨTC . You should always examine the solution from ΨTC to be certain that you have converged to the stable steady-state solution you want. If you have not, reducing δ_0 will solve the problem if the initial value problem has a steady-state solution. If you are applying ΨTC to a problem with no steady-state solution from your initial data, ΨTC will not produce anything useful.

2.5.4 ▪ Sparse Jacobians

If $\mathbf{F}'$ is sparse, you may have the option to compute a sparse factorization without pivoting. If, for example, $\mathbf{F}'$ is symmetric and positive definite, this is the way to proceed. For general $\mathbf{F}'$, however, pivoting can be essential for a factorization to produce useful solutions. For sparse problems, the cost of pivoting can be large and it is tempting to avoid it. If line search fails and you have disabled pivoting in your sparse factorization, it's probably a good idea to reenable it and find a way to deal with the storage costs.

The cost estimates for a difference Jacobian change if F' is sparse, as does the cost of the factorization. In the sparse case one can compute several columns of the Jacobian with a single new function evaluation. The methods for doing this for general sparsity patterns [37, 41] are too complex for this book. The **SparseDiffTools.jl** [160] package in Julia is a way to incorporate sparse differences into your Jacobian evaluation function.

Even though sparse differencing can be done efficiently, sparse matrix factorization is still problematic. The factors are typically more dense than the original matrix, a phenomenon called **fill-in**. Hence a sparse LU factorization cannot be expected to overwrite the original matrix and to use little new storage beyond that. One can do a symbolic factorization [45] and figure out how much storage is needed and then preallocate that storage. We challenge the reader to do that in one of the projects at the end of this chapter.

Our **nsol.jl** code does a sparse LU factorization by default if F' is sparse and pays the price for the memory allocation. One cannot expect to overwrite the storage for F' with the factors and must use the output of a symbolic factorization to avoid allocating storage many times. Banded matrices [7, 111] are an exception, and we will look at an example in section 2.7.3.

Finally, the general sparse solvers in Julia are based on the SparseSuite codes [45] which require double precision. Hence one cannot do the linear algebra in single precision in this case. Some matrices with special structure, such as banded matrices, do allow this, but the advantages are not as compelling as in the dense case.

2.6 ▪ Precomputed and Preallocated Data

Many of the examples in this book depend on data that one should compute and store only once. Our solvers use the kwarg **pdata** for this and allow you to pass that to the function, the Jacobian, and the Jacobian- (and preconditioner-)vector products.

More importantly, allocation of data is expensive in Julia, and our codes use in-place solvers like **lu!** when possible. The solvers also require that you preallocate space for the Jacobian and residual. Doing this is easy and the examples will illustrate the process.

2.7 ▪ Five Nonlinear Systems

In this section we show how to use **nsol.jl** for five problems. The computations in this section use very few of the options in the solver and show how to use **nsol.jl** in the most simple manner. In section 2.8.1 we will explain the options for **nsol.jl** in detail and revisit the examples in this section. In this section we will use Newton's method (sham=1) for the five examples. The default for systems is sham=5, and we will show why that is a good idea in section 2.8.1. The author of this book also likes sham=Inf, which differs from the chord method in that the Jacobian is updated if the reduction in the residual norm is larger than **resdec** (default = 0.1). The reader should look at [21] for one vintage view of how often one should update the Jacobian.

The interface to **nsol.jl** is very similar to that of the scalar code **nsolsc.jl** from the previous chapter. The most significant differences are in the way you must manage storage. You must allocate storage for your function and Jacobian in the calling program, and the Julia functions you write for evaluating functions and Jacobians will overwrite that storage.

We will call **nsol.jl** with mostly the default options:

```
nsolout=nsol(F!, x0, FS, FPS, J!; sham =1)
```

Here **nsolout** is the output data structure, which, as we said above, is very similar to the one for our scalar codes. x_0 is the initial iterate. The new thing is the preallocated storage for the function **FS** and Jacobian **FPS**. **F!** is the function for evaluation the residual and **J!** the function for evaluating the Jacobian. The syntax for **F!** is

```
FS=F!(FS,x)
```

F! will return **FS = F(x)**, overwriting the data stored in the array **FS**. Similarly one computes the Jacobian with

```
FP=J!(FS,FPS,x)
```

to compute **FPS = F′(x)**. Note that **FS** must also be an argument for the Jacobian evaluation function **J!**. The reason for this, and the reason that **J!** is last in the argument list, is that **J!** will default to a finite difference Jacobian evaluation if left out of the call to **nsol.jl**. So

```
nsolout=nsol(F!, x0, FS, FPS; sham=1)
```

will solve $\mathbf{F}(\mathbf{x}) = 0$ with Newton's method and a finite difference Jacobian.

2.7.1 ▪ A Simple Two-Dimensional Example

This example is from [54]. Here $N = 2$ and

$$\mathbf{F}(\mathbf{x}) = \left(\begin{array}{c} x_1^2 + x_2^2 - 2 \\ \exp(x_1 - 1) + x_2^2 - 2 \end{array} \right).$$

This function is simple enough for us to put the Julia code from **src/TestProblems/Systems/simple!.jl** that computes the function here:

```
function simple!(FV,x)
    FV[1]=x[1]*x[1] + x[2]*x[2] -2.0;
    FV[2]=exp(x[1]-1) + x[2]*x[2] - 2.0;
    return FV
end
```

In the experiment we report in this section, take the defaults $\tau_a = 10^{-12}$ and $\tau_r = 10^{-6}$ and use a finite difference Jacobian. We investigated two initial iterates. For $\mathbf{x}_0 = (2, 0.5)^T$, the step length was reduced twice on the first iteration. Full steps were taken after that. This is an interesting example because the iteration can stagnate at a point where $\mathbf{F}'(\mathbf{x})$ is singular. If $\mathbf{x}_0 = (3,5)^T$, the line search will fail and the stagnation point will be near the x_1 axis, where the Jacobian is singular.

The code for the experiment is included in the notebook in

NotebookSIAMFANL/src/Chapter2/TwoDexample.jl

We will list the part of **TwoDexample.jl** that calls the solver and let the interested reader look at the source for the details of the plotting. Note how I've allocated the storage for the residual and the Jacobian:

```
function TwoDexample()
   x0a=[2.0, .5];
   x0b=[3.0, 5.0];
# Allocate some room for the residual and Jacobian
   FS=zeros(2,);
   FPS=zeros(2,2);
   nouta=nsol(simple!, x0a, FS, FPS; sham=1, keepsolhist=true);
   xa=nouta.solhist[1,:];
   ya=nouta.solhist[2,:];
   noutb=nsol(simple!, x0b, FS, FPS; sham=1, keepsolhist=true);
   xb=noutb.solhist[1,:];
   yb=noutb.solhist[2,:];
#
# Some plotting commands you may not want to look at are here.
#
end
```

Note that we ask for the solution history in each call to **nsol.jl** because we plot that history in the figure. In Figure 2.1 we plot the iteration history for both choices of initial iterate on a contour plot of $\|\mathbf{F}\|$. The iteration that stagnates converges, but not to a root! Line search codes that terminate when the step is small should also check that the solution is an approximate root, perhaps by evaluating $\mathbf{F}$ (see section 1.9.2).

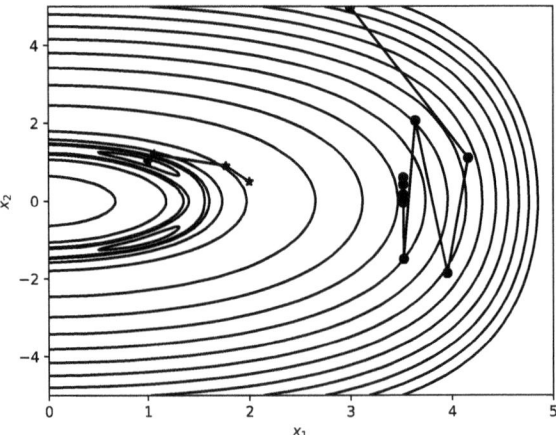

Figure 2.1: Solution of two-dimensional example with **nsol.jl**.

2.7.2 ▪ Chandrasekhar H-Equation

The Chandrasekhar H-equation [27, 32] is

$$\mathcal{F}(H)(\mu) = H(\mu) - \left(1 - \frac{c}{2}\int_0^1 \frac{\mu H(\nu)\,d\nu}{\mu + \nu}\right)^{-1} = 0. \tag{2.3}$$

This equation arises in radiative transfer.

This equation has a well-understood dependence on the parameter c [48,141]. The equation has unique solutions at $c = 0$ and $c = 1$ and two solutions for $0 < c < 1$. There is a simple fold singularity [106] at $c = 1$. Only one [27, 32] of the two solutions for $0 < c < 1$ is of physical interest, and that is the one easiest to find numerically. One must perform a continuation computation to find the other one, which we do in section 5.2.

The structure of the singularity is preserved if one discretizes the integral with any quadrature rule with positive weights that integrates constants exactly. This fact enables one to readily understand the behavior of nonlinear solvers.

We will approximate the integrals by the composite midpoint rule:

$$\int_0^1 f(\mu)\,d\mu \approx \frac{1}{N}\sum_{j=1}^{N} f(\mu_j),$$

where $\mu_i = (i - 1/2)/N$ for $1 \le i \le N$. The resulting discrete problem is

$$\mathbf{F}(\mathbf{x})_i = x_i - \left(1 - \frac{c}{2N}\sum_{j=1}^{N}\frac{\mu_i x_j}{\mu_i + \mu_j}\right)^{-1}. \tag{2.4}$$

As is the case with most integral equations, increasing N has no effect on the conditioning of the Jacobian nor on the iteration statistics [2, 119]. Hence we should observe that the iteration statistics are independent of N.

We will use the Fourier transform approach from [112] and [113] rather than the matrix-based approach from [111]. One can simplify the approximate integral operator in (2.4) and expose some useful structure. Since

$$\frac{c}{2N}\sum_{j=1}^{N}\frac{x_j \mu_i}{\mu_j + \mu_i} = \frac{c(i - 1/2)}{2N}\sum_{j=1}^{N}\frac{x_j}{i + j - 1},$$

the approximate integral operator is the product of a diagonal matrix and a Hankel matrix, and one can use a fast Fourier transform (FFT) to evaluate the operator-vector product with $O(N\log(N))$ work [75, 112]. We use the Julia package **FFTW.jl** [102]. The documentation for that package and the paper [67] are very much worth reading.

We can express the approximation of the integral operator in matrix form

$$\mathbf{M}(\mathbf{x})_{ij} = \frac{c(i - 1/2)}{2N}\sum_{j=1}^{N}\frac{x_j}{i + j - 1}$$

and compute the Jacobian analytically as

$$\mathbf{F}'(\mathbf{x}) = \mathbf{I} - \mathrm{diag}(\mathbf{G}(\mathbf{x}))^2 \mathbf{M},$$

where

$$\mathbf{G}(\mathbf{x})_i = \left(1 - \frac{c}{2N} \sum_{j=1}^{N} \frac{x_j \mu_i}{\mu_j + \mu_i}\right)^{-1}.$$

Hence the data for the Jacobian is already available after one computes $\mathbf{F}(\mathbf{x}) = \mathbf{x} - \mathbf{G}(\mathbf{x})$ and the Jacobian can be computed with $O(N^2)$ work. We do that in this example, and therefore the only part of the solve that requires $O(N^3)$ work is the matrix factorization. Of course, one can also approximate the Jacobian with finite differences at a cost of $O(N^2 \log(N))$ work.

This is the first example of the need for **precomputed data**. The file **Hequation.jl** in **Test-Problems/Systems** has not only the functions you need to evaluate the nonlinear residual and Jacobian, but also the function **heqinit**, which initializes the precomputed data. These data include the following:

- The output from **plan_fft!**, which initializes the data for the FFT. Note the **!**; this indicates that the FFT will overwrite its input and save many allocations.

- Vectors for the quadrature points and intermediate storage for the Hankel matrix product.

- The value of c, which you may change with the **setc!** function.

We do not list the lengthy source code for the module **Hequation.jl**.

The precomputed data saves a lot of time and keeps the code short and organized. This book, like [107, 111], strongly advocates the use of precomputed data for linear and nonlinear solvers. All of our solvers for systems of equations support the use of precomputed data.

The Julia code **HeqSolutions.jl** in the notebook solves and plots (see Figure 2.2) the H-equation with initial iterate $x_0 = (1, \ldots, 1)^T$, $\tau_a = \tau_r = 10^{-10}$, $N = 100$. We compare the solutions with $c = 0.5$ and $c = 0.9$. We use a finite difference Jacobian in this example. We will do a more detailed comparison with an analytic Jacobian and single precision linear algebra in section 2.8.1. The kwargs for the tolerances are the same as in the scalar codes. The new thing is the kwarg for the precomputed data. The keyword is **pdata**.

```
"""
HeqSolutions()
Draw Fig 2.2
"""
function HeqSolutions()
    n = 100
    c = 0.9
    x0 = ones(n)
    #
    #   Build the structure with the precomputed data.
    #
    hdata = heqinit(x0, c)
    #
    #   Allocate the room for the residual and Jacobian.
    #
```

```
    FS = ones(n)
    FPS = ones(n, n)
    #
    #    Call the solver twice.
    #
    nsolout9 = nsol(heqf!, x0, FS, FPS;
              rtol = 1.e-10, atol = 1.e-10, pdata = hdata, sham=1)
    #    Change c to .5
    setc!(hdata,.5)
    nsolout5 = nsol(heqf!, x0, FS, FPS;
              rtol = 1.e-10, atol = 1.e-10, pdata = hdata, sham=1)
    #
    # Showtime! Make the plot.
    #
    axis([0, 1.0, 1.0, 1.9])
    mu = hdata.mu
    H9 = nsolout9.solution
    H5 = nsolout5.solution
    plot(mu, H9, "k-", mu, H5, "k--")
    xlabel(L"\mu")
    ylabel(L"H(\mu)")
    legend(["c=.9", "c=.5"])
end
```

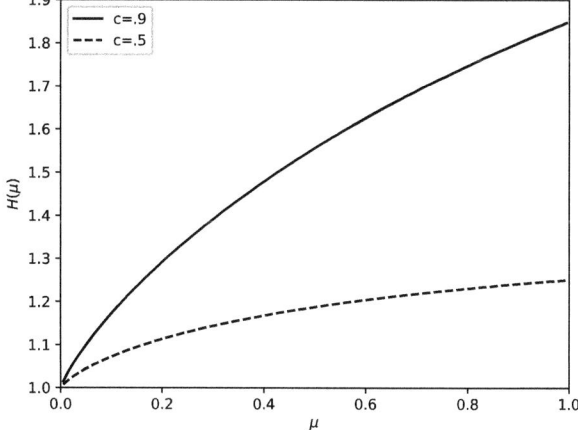

Figure 2.2: Solution of the H-equation for $c = 0.5, 0.9$.

2.7.3 ▪ A Two-Point Boundary Value Problem

This example, an exercise from [10], has a **banded Jacobian**. A banded matrix only has nonzero elements on bands about the diagonal. So **A** is banded with upper bandwidth m_u and lower bandwidth m_l if $\mathbf{A}_{ij} = 0$ unless

$$i - m_l \le j \le m_u + i.$$

An LU or QR factorization of **A** can reuse the storage allocated for **A** if one pads **A** with two extra upper bands. We do that in this application and use the QR factorization

```
qr!(A)
```

The reason we use **qr** and not **lu** for banded matrices is that **qr**, as of the date this book was written, supports our mixed precision approach of doing linear algebra in single precision and **lu** does not. This is likely to change.

We can use the fast band solvers from LAPACK and Julia via the Julia package **Banded-Matrices.jl** [144]. In this example it is easy to compute the Jacobian analytically. Most of the Jacobian is independent of the nonlinear iteration, and we can store the data for that in the precomputed data.

We seek $v \in C^2([0, 20])$ such that

$$v''(t) + (4/t)v'(t) + (tv(t) - 1)v(t) = 0; v'(0) = v(20) = 0. \tag{2.5}$$

This problem has at least two solutions. One, $v = 0$, is not interesting, so the objective is to find a nonzero solution.

We begin by converting (2.5) to a first-order system for

$$U = \begin{pmatrix} u_1 \\ u_2 \end{pmatrix} = \begin{pmatrix} v \\ v' \end{pmatrix}.$$

The equation for U is

$$U'(t) = \begin{pmatrix} u_1(t) \\ u_2(t) \end{pmatrix}' = G(t, U(t)) = \begin{pmatrix} u_2(t) \\ -(4/t)u_2(t) - (tu_1(t) - 1)u_1(t) \end{pmatrix}. \tag{2.6}$$

We will discretize the problem on an equally spaced mesh $\{t_i\}_{i=1}^N$, where $t_i = (i-1)dt$ and $dt = 20/(N-1)$. Letting $U_i \approx U(t_i)$, we can express the problem for $\{U_i\}_{i=1}^N$ as a nonlinear equation $\mathbf{F}(\mathbf{x}) = 0$ with a banded Jacobian of upper and lower bandwidth two by grouping the unknowns at the same point on the mesh:

$$\mathbf{x} = (U_1^T, U_2^T, \ldots, U_N^T)^T.$$

In this way $x_{2i+1} \approx v(t_i)$ and $x_{2i} \approx v'(t_i)$. The boundary conditions are the first and last equations

$$\mathbf{F}(\mathbf{x})_1 = x_2 = 0 \quad \text{and} \quad \mathbf{F}(\mathbf{x})_{2N} = x_{2N-1} = 0.$$

$u_1' = u_2$ is expressed in the odd components of F as

$$\mathbf{F}(\mathbf{x})_{2i+1} = x_{2i+1} - x_{2i-1} - (h/2)(x_{2i} + x_{2i+2})$$

for $1 \le i \le N - 1$.

The even components of F are the discretization of the original differential equation

$$\mathbf{F}(\mathbf{x})_{2i} = x_{2i+2} - x_{2i} + (h/2)(\Phi_{i+1}(\mathbf{x}) + \Phi_i(\mathbf{x})), \quad 1 \le i \le N - 1.$$

Here

$$\Phi_i(x) = (4t_i^\dagger)x_{2i} + (t_i x_{2i-1} - 1)x_{2i-1}$$

and

$$t^\dagger = \begin{cases} 1/t & \text{if } t > 0, \\ 0 & \text{if } t = 0. \end{cases}$$

If we organize $\mathbf{x}$ in this way, we can split $\mathbf{F}$ into linear and nonlinear parts:

$$\mathbf{F}(\mathbf{x}) = \mathbf{D}\mathbf{x} + \mathbf{P}(\mathbf{x}). \tag{2.7}$$

In (2.7)

$$(\mathbf{D}\mathbf{x})_1 = x_2 \quad \text{and} \quad (\mathbf{D}\mathbf{x})_{2N} = x_{2N-1},$$

and for $i = 1, \ldots, N-1$

$$\begin{aligned} (\mathbf{D}\mathbf{x})_{2i+1} &= x_{2i+1} - x_{2i-1} - (h/2)(x_{2i} + x_{2i+2}), \\ (\mathbf{D}\mathbf{x})_{2i} &= x_{2i+2} - x_{2i}. \end{aligned}$$

The nonlinear term is

$$\begin{aligned} (\mathbf{P}(\mathbf{x}))_{2i+1} &= 0 & \text{for } i = 0, \ldots, N, \\ (\mathbf{P}(\mathbf{x}))_{2i} &= (h/2)(\Phi_{i+1}(\mathbf{x}) + \Phi_i(\mathbf{x})) & \text{for } i = 0, \ldots, N-1. \end{aligned}$$

The Jacobian is **banded**. It is easy to see from the formula for $\mathbf{F}$ that $\mathbf{F}'$ is banded with $m_u = m_l = 2$. Moreover, only the even rows of $\mathbf{F}'$ depend on $\mathbf{x}$. Hence one can store $\mathbf{D}$ in the precomputed data and only have to differentiate $\mathbf{P}$ with each iteration.

The astute reader will notice (or remember) that the function $U \equiv 0$ is a solution of (2.6). If you change the initial iterate in our example, you may well converge to the zero solution. We can find a nonzero solution using the initial iterate

$$v(t) = e^{-t^2/10} \quad \text{and} \quad v'(t) = -te^{-t^2/10}/5.$$

The solver struggles, with the line search being active for three of the nine iterations. We plot that solution in Figure 2.3.

The file **Fvpb!.jl** in **TestProblems/Systems** contains the evaluation of the function and Jacobian for this problem as well as a function to initialize the precomputed data. We store the Jacobian as a banded matrix and use the **BandedMatrices.jl** [144] package. The Julia code **BVPsolution.jl** in the notebook solves the problem with $N = 801$ (so $h = 1/40$) and creates the plot. The core function within **BVPsolution.jl** is bvp_solve from the examples. **bvp_solve.jl** builds the initial iterate and solves the problem. The plotting commands are very similar to those we use for the H-equation problem in section 2.7.2, and we omit those commands in the listing for **bvp_solve**:

```
"""
bvp_solve

Solve the boundary value problem to make Figure 2.3.
"""
function bvp_solve(n = 801, T = Float64)
    # set it up
    bdata = bvpinit(n, T)
    #
    U0 = zeros(2n)
    FV = zeros(2n)
    # BandedMatrix with the correct number of bands.
```

```
FPV = bdata.JacS
#
# tv = 0:h:20 is the spatial mesh
#
tv = bdata.tv
#
# Build the initial iterate
#
sv = -.1 * tv .* tv
v = exp.(sv)
vp = -.2 * tv .* v
U0[1:2:2n-1] = v
U0[2:2:2n] = vp
#
# Call the solver. Jbvp! is my analytic Jacobian evaluation.
#
bvpout = nsol(Fbvp!, U0, FV, FPV, Jbvp!;
               rtol = 1.e-10, pdata = bdata, sham=1, jfact=qr!)
return (bvpout = bvpout, tv = tv)
end
```

The function **bvpinit** allocates storage for the Jacobian as a matrix with lower bandwidth 2 and upper bandwidth 4:

```
FVP = BandedMatrix{T}(Zeros(2n, 2n), (2, 4))
```

Here T is the precision for storing the Jacobian. So if the call to bvp_solve is

```
bvp_solve(801, Float32)
```

the Jacobian will be stored in single precision and the factorization and linear solve will be done in single precision. The solver **nsol.jl** uses **qr** to factor the Jacobian by default if the matrix is banded. If you allocate the extra storage (and you should), then set the kwarg **jfact** to **qr!** as we did in **bvp_solve**.

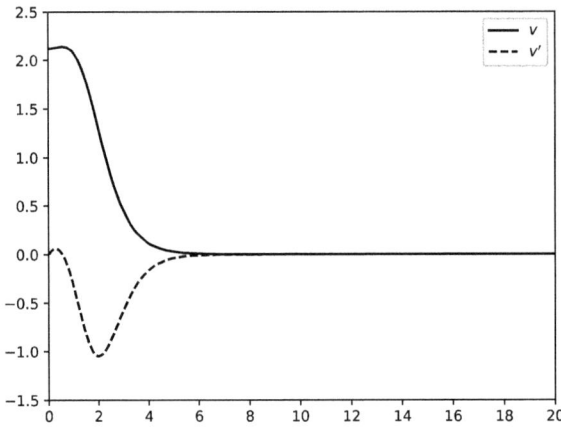

Figure 2.3: Solution (2.5).

2.7.4 ▪ Stiff Initial Value Problems and the Buckling Beam

We will consider the problem for a beam under load (see [106] and Chapter 7 of [135]). We apply ΨTC to this problem in section 2.8.6. The time-dependent problem is

$$u_t = u_{xx} + \lambda \sin(u), \qquad (2.8)$$

$$u(0,t) = u(1,t) = 0 \quad \text{for all } 0 < t \leq 1$$

and initial datum

$$u(x,0) = x(1-x)(2-x)e^{-10x(1-x)(2-x)} \quad \text{for all } 0 \leq x \leq 1.$$

The initial datum $u_0 \geq 0$ and, for $\lambda > \pi^2$, is close to the unstable steady state $u \equiv 0$. As you will see in section 2.8.6, we need to think about the parameter λ to get something interesting. The choice $\lambda = 20$ will work fine.

The steady-state problem

$$u_{xx} + \lambda \sin(u), \quad u(0) = u(1) = 0 \qquad (2.9)$$

is another example of a pitchfork bifurcation. Figure 2.4, taken from Chapter 7 of [135], illustrates the application and the pitchfork far more vividly than any computer-generated figure could.

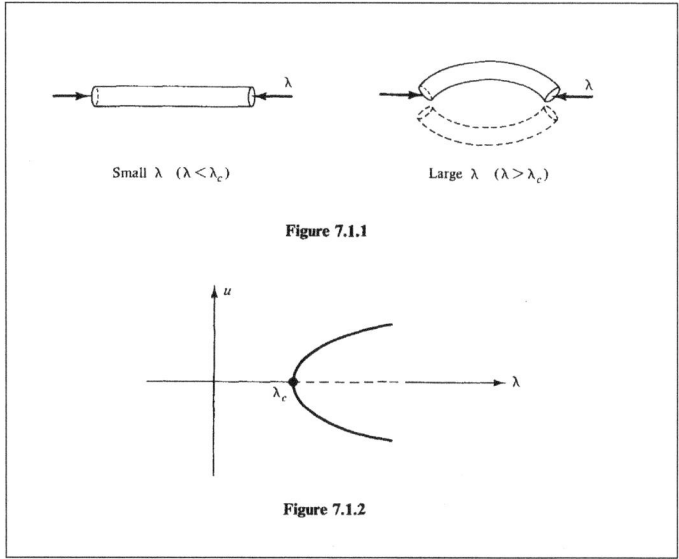

Figure 2.4: Classic illustration for the buckling beam. Reprinted with permission from [135].

Note that the point $\lambda_c = \pi^2$ is where the bifurcation takes place.

Time-accurate integration

Nonlinear solvers are important parts of codes for **stiff initial value problems**. In general terms [10, 176, 177], stiffness means that either implicit methods must be used to integrate in time or, in the case of an explicit method, very small time steps must be taken.

If the problem is nonlinear, a nonlinear solver must be used at each time step. The most elementary example is the implicit Euler method. To solve the initial value problem

$$\mathbf{u}' = \mathbf{G}(\mathbf{u}), \mathbf{u}(0) = \mathbf{u}^0 \qquad (2.10)$$

with the implicit Euler method, we specify a time step δ_t and approximate the value of the solution at the mesh point $n\delta_t$ by $\mathbf{u}^n$, where $\mathbf{u}^n$ solves the nonlinear equation

$$\mathbf{u}^n = \mathbf{u}^{n-1} + \delta_t \mathbf{G}(\mathbf{u}^n). \qquad (2.11)$$

The nonlinear solver is given the function

$$\mathbf{F}(\mathbf{w}) = \mathbf{w} - \mathbf{u}^{n-1} - \delta_t \mathbf{G}(\mathbf{w})$$

and an initial iterate. The initial iterate is usually either $\mathbf{w}_0 = \mathbf{u}^{n-1}$ or a linear predictor $\mathbf{w}_0 = 2\mathbf{u}^{n-1} - \mathbf{u}^{n-2}$. In many codes [10, 20, 162] the termination criterion is based on small Newton step lengths, usually something like (1.19). This eliminates the need to evaluate the function only to verify a termination condition. Similarly, the Jacobian is updated very infrequently—rarely at every time step and certainly not at every nonlinear iteration. This combination can lead to problems if the Jacobian is ill-conditioned and varies rapidly with time [118], but is usually very robust. The time step δ_t depends on n in any modern initial value problem code. Hence the solver sees a different function (varying $\mathbf{u}^{n-1}$ and δ_t) at each time step. We refer the reader to the literature [10, 176, 177] for a complete account of how nonlinear solvers are managed in initial value problem codes and focus here on a very basic example. Here we update the Jacobian at each time step and use the chord method for the nonlinear iteration at each time step.

We solve the initial value problem for the buckling beam (2.8) on a spatial mesh with width $\delta_x = 1/64$ and use a time step of $\delta_t = 0.01$. The unknowns are approximations to $u(x_i, t_n)$ for the interior nodes $\{x_i\}_{i=1}^{63} = \{i\delta_x\}_{i=1}^{63}$ and times $\{t_i\}_{i=1}^{200} = \{i\delta_t\}_{i=1}^{200}$. Our discretization in space is the standard central difference approximation to the second derivative with homogeneous Dirichlet boundary conditions. The discretized problem is a stiff system of 63 ordinary differential equations. Typically implicit methods are used for such systems [10, 176, 177]. The simple backward Euler method well illustrates the issues for nonlinear solvers.

For a given time step n and time increment δ_t, the components of the function $\mathbf{F}$ sent to **nsol.jl** are given by

$$(\mathbf{F}(\mathbf{w}))_i = w_i - u_i^{n-1} + \delta_t(\lambda \sin(w_i) + (\mathbf{D}_2 \mathbf{w})_i)$$

for $1 \le i \le N = 63$. The discrete second derivative $\mathbf{D}_2$ is the tridiagonal matrix with -2 along the diagonal and 1 along the sub- and superdiagonals.

This problem has the same structure as the boundary problem in section 2.7.3. The Jacobian is the sum of a linear operator that is independent of $\mathbf{w}$ and a nonlinear part. The nonlinear part of this problem is particularly simple, a substitution operator, and has a diagonal Jacobian.

In Figure 2.5 we plot the temporal history of the integration. As time increases, $u(x, t)$ approaches the stable steady-state solution. The function **beamtimedep.jl** from **Notebook/ src/Chapter2** generated the figure. **beamtimedep.jl** uses the code for the example problem **src/Examples/ivpBeam.jl**. At the end of the integration the norm of the residual was $\approx 2 \times 10^{-9}$.

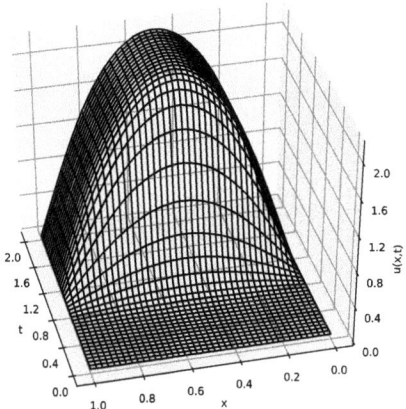

Figure 2.5: Temporal integration for the buckling beam.

2.7.5 ▪ Convection-Diffusion Equation

This is an example from [107, 111] of a semilinear (i.e., linear in the highest order derivative) elliptic partial differential equation (PDE). This will show how to use the general sparse matrix data structure in Julia.

The problem is

$$- \nabla^2 u + 20u(u_x + u_y) = f \qquad (2.12)$$

with homogeneous Dirichlet boundary conditions

$$u(x, 1) = u(x, 0) = u(0, y) = u(1, y) = 0$$

on the unit square $(0, 1) \times (0, 1)$. Here ∇^2 is the Laplacian operator

$$\nabla^2 = \frac{\partial^2}{\partial^2 x} + \frac{\partial^2}{\partial^2 y}.$$

f has been constructed so that the exact solution is the discretization of

$$u^*(x, y) = 10xy(1 - x)(1 - y) \exp(x^{4.5}).$$

We discretized on a uniform mesh with 31 interior grid points in each direction using centered differences and terminated with

$$\tau_a = 10^{-10} \quad \text{and} \quad \tau_r = 10^{-7}.$$

The reader will get a wonderful opportunity to experiment with finer grids in the notebook and in Chapter 3.

The physical grid is two-dimensional, but solvers expect one-dimensional vectors. One can resolve this by ordering the unknowns and then forming the matrix representations of the linear transformations. Julia's sparse matrix data structure and sparse solvers [44, 45] from **SuiteSparse.jl** make this easy. Use the Julia `reshape` command to move between the two-dimensional grid, which is useful for defining coefficient functions and plotting the results, to the one-dimensional grid for the solvers.

So if the two-dimensional grid is $n \times n$, then the spatial mesh width is $h = 1/(n+1)$ and the grid points are (x_i, x_j), where $x_i = i * h$ for $1 \leq i \leq n$. There are $N = n^2$ unknowns and I will order the grid points by varying the x coordinate before the y coordinate. This means that if the two-dimensional data are $\mathbf{u}^{2D}$, an $n \times n$ array where

$$\mathbf{u}_{i,j}^{2D} \approx u(x_i, x_j),$$

then the vector one sends to the solvers uses the same storage and simply is arranged as a vector $\mathbf{u}$. The command for this is

```
U=reshape(U2D,N)
```

The contents of the storage are unchanged and one can refer to

```
U2D
```

as two-dimensional data and to

```
U
```

as one-dimensional data without doing anything more than the reshape to map the two-dimensional representation to the one-dimensional one.

Julia stores two-dimensional arrays by columns. Our ordering means that discretization of the partial derivative with respect to x is sparse with the only nonzeros on the sub- and superdiagonals. Similarly the partial derivative with respect to y is sparse with the only nonzeros separated from the diagonal by n. The sparse matrix for the discrete Laplacian has five nonzero bands, the diagonal and the nonzero bands used by the first partial derivatives. We use the Julia `spdiagm` command to build these matrices. For example, the code for the negative Laplacian is

```
"""
Lap2d(n)

returns the negative Laplacian in two space dimensions
on n x n grid.

Unit square, homogeneous Dirichlet BC
"""
```

```
function Lap2d(n)
# hm2=1/h^2
hm2=(n+1.0)^2;
maindiag=fill(4*hm2,(n^2,));
sxdiag=fill(-hm2,(n^2-1,));
sydiag=fill(-hm2,(n^2-n,));
for iz=n:n:n^2-1
    sxdiag[iz]=0.0;
end
D2=spdiagm(-n => sydiag, -1 => sxdiag, 0=> maindiag,
            1 => sxdiag, n => sydiag);
return D2
end
```

One builds the sparse matrices $\mathbf{D}_x$ for $\partial/\partial x$ and $\mathbf{D}_y$ for $\partial/\partial y$ in a similar manner. The codes for these operators, which we will use many times in this book, are in the file **PDE_Tools.jl** in the **src/TestProblems/Systems** directory.

One can express the discrete problem in terms of the (precomputed!) sparse matrices for the differential operators as

$$\mathbf{F}(\mathbf{u}) = \mathbf{D}_2\mathbf{u} + 20\mathbf{u} \cdot (\mathbf{D}_x + \mathbf{D}_y)\mathbf{u} - \mathbf{f}, \qquad (2.13)$$

where the $\cdot$ denotes componentwise multiplication. Using the very useful formula

$$\mathbf{F}'(\mathbf{u})\mathbf{w} = \left.\frac{d}{d\epsilon}\mathbf{F}(\mathbf{u} + \epsilon\mathbf{w})\right|_{\epsilon=0}, \qquad (2.14)$$

it is easy to see that

$$\mathbf{F}'(\mathbf{u})\mathbf{w} = \mathbf{D}_2\mathbf{w} + 20\mathbf{w} \cdot (\mathbf{D}_x + \mathbf{D}_y)\mathbf{u} + 20\mathbf{u} \cdot (\mathbf{D}_x + \mathbf{D}_y)\mathbf{w}$$

and therefore

$$\mathbf{F}'(\mathbf{u}) = \mathbf{D}_2 + 20\mathrm{diag}(\mathbf{u})(\mathbf{D}_x + \mathbf{D}_y) + 20\mathrm{diag}((\mathbf{D}_x + \mathbf{D}_y)(\mathbf{u})).$$

The analytic Jacobian is sparse with five nonzero diagonals. When we call **nsol.jl** and allocate storage for the sparse Jacobian, **nsol.jl** will use the sparse solvers, with the sparse LU factorization as the default.

We store $\mathbf{D}_2$, $\mathbf{CV} = (\mathbf{D}_x + \mathbf{D}_y)$, and the right-hand side $rhs = \mathbf{f}$ in the structure `pdata` of precomputed data. The residual evaluation function (in **src/TestProblems/Systems/EllipticPDE.jl**) is quite simple:

```
"""
pdeF!(FV, u, pdata)

Residual using sparse matrix-vector multiplication
"""
function pdeF!(FV, u, pdata)
D2=pdata.D2
CV=pdata.CV
rhs=pdata.RHS
FV .= D2*u + 20.0*u.*(CV*u) - rhs
end
```

The Jacobian evaluation is tricky because one must think about how to assemble the sparse matrix in order to avoid expensive allocations. We invite the reader to look at the function **pdeJ!** in **src/TestProblems/Systems/EllipticPDE.jl** to see.

In Figure 2.6 we compare the residual histories for Newton's method, which does a sparse matrix factorization at every iteration, and the default Shamanskii method with $m = 5$. The Shamanskii iteration evaluates and factors the Jacobian only twice and takes only two more nonlinear iterations than Newton. We will see in the notebook how large those savings can be in the context of a larger example of this problem.

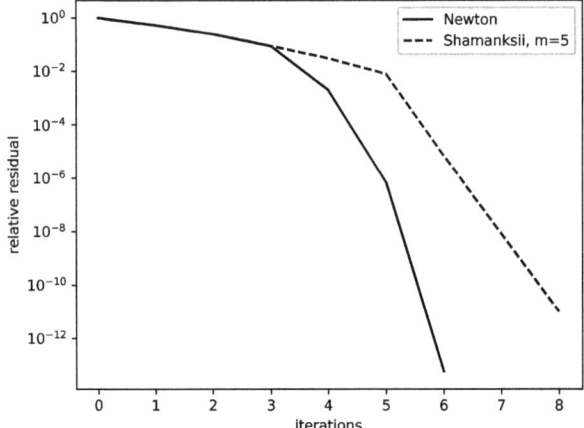

Figure 2.6: Residual history for the elliptic PDE.

2.7.6 ▪ Pseudo-transient Continuation and the Buckling Beam

ΨTC confronts the same issues with direct solvers as the Newton–Armijo iteration, and the ideas in section 2.1 are equally applicable. Preallocation of storage, exploitation of structure, and mixed precision are all important for ΨTC .

In this section we will apply ΨTC to the buckling beam problem (2.9). The code **ptcsol.jl** requires preallocation of storage for the function and Jacobian in the same way that **nsol.jl** does. We initialized ΨTC with $\delta_0 = 0.01$.

Our example problem **ptcBeam.jl** uses the same initial data and spatial grid that we used for the time-dependent problem (2.8). The code compares the results for ΨTC to Newton's method for the same data. As you should expect, the ΨTC iteration converges to the positive stable steady state and the Newton iteration converges to the unstable zero solution. We use an analytic Jacobian. Our residual function is **FBeam!.jl** and the Jacobian is **BeamJ!.jl**. These functions are in the file

src/TestProblems/Systems/FBeam!.jl.

The precomputed data include the tridiagonal discrete Laplacian and the grid points. The calls to **nsol.jl** and **ptcsol.jl** within **src/Examples/ptcBeam.jl** are

```
bout=ptcsol(FBeam!, u0, FS, FPS, BeamJ!;
            rtol=1.e-10, pdata=bdata, delta0=delta, maxit=maxit);
qout=nsol(FBeam!, u0, FS, FPS, BeamJ!; pdata=bdata, sham=1);
```

Note the similarities in the calling sequences.

We plot the residual histories in Figure 2.7. ΨTC converged to the stable nonnegative steady state in 24 steps with a final residual norm of $\approx 7 \times 10^{-12}$. This is a significant improvement over the direct integration to steady state in the previous section. The Newton iteration converged to the unstable solution in only two iterations. As you can see, the residuals are not monotone decreasing for ΨTC . You should expect this because the iteration must pass through transients before converging to steady state.

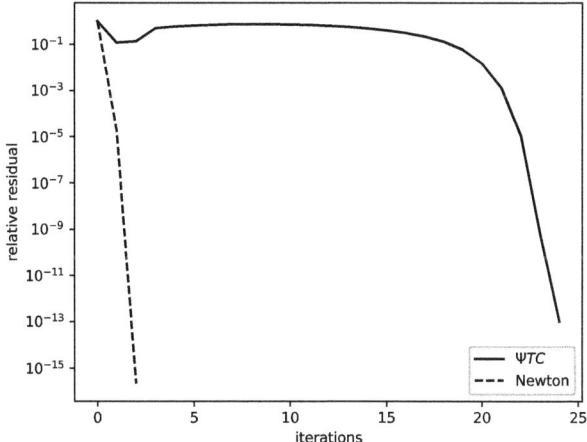

Figure 2.7: ΨTC residual histories for the buckling beam.

Our choices of $\delta_0 = 0.01$ and the spatial mesh width of $\delta_x = 1/64$ are not independent. The Rosenbrock method is explicit and therefore will become unstable if the initial pseudo-time step is too large. In the context of ΨTC this means that, for a fixed δ_x, there is a limit to how large δ_0 can be if one wants to be certain to converge to the correct steady-state solution $\mathbf{u}^* = \lim_{t\to\infty} \mathbf{u}(t)$ of the initial value problem. The condition that δ_0 be sufficiently small is serious. It should be clear to you that a choice of δ_0 that is extremely large will make the ΨTC iteration close to Newton's method and thereby converge to the nearest solution, stable or not.

A more subtle problem is that choice of δ_0 that is a bit too large may cause the first step or two of Rosenbrock to leave the trajectory for the stable steady state of interest. In that case the SER update for δ will correct the stability, but you may then converge to a different steady state. You can explore this in the project in section 2.9.6.

2.8 ▪ Notebook: Solvers for Chapter 2

We will follow the pattern of Chapter 1 and present two solvers: a Newton code and a ΨTC code. Both codes are for systems of equations and use direct methods to compute the step. We returned the solution history for the simple two-dimensional example in section 2.6, but will not do that again.

2.8.1 ▪ nsol.jl

nsol.jl solves systems of nonlinear equations and computes the Newton step with direct linear solvers. Let's look at the docstrings.

[2]: `?nsol`

[2]:
```
nsol(F!, x0, FS, FPS, J!=diffjac!; rtol=1.e-6, atol=1.e-12,
        maxit=20, solver="newton", sham=5, armmax=10,
        resdec=.1, dx = 1.e-7, armfix=false,
        pdata = nothing, jfact = klfact, printerr = true,
        keepsolhist = false, stagnationok=false)
```

C. T. Kelley, 2022

Julia versions of the nonlinear solvers from my SIAM books. Herewith: nsol

You must allocate storage for the function and Jacobian in advance -> in the calling program <- ie. in FS and FPS

Inputs:

- F!: function evaluation, the ! indicates that F! overwrites FS, your preallocated storage for the function.

  So FS=F!(FS,x) or FS=F!(FS,x,pdata) returns FS=F(x)

  Your function MUST have -> return FS <- at the end. See the examples in the docstrings and in TestProblems/Systems/simple.jl

- x0: initial iterate

- FS: Preallocated storage for function. It is a vector of size N

  You should store it as (N) and design F! to use vectors of size (N). If you use (N,1) consistently instead, the solvers may work, but I make no guarantees.

- FPS: preallocated storage for Jacobian. It is an N x N matrix

- J!: Jacobian evaluation, the ! indicates that J! overwrites FPS, your preallocated storage for the Jacobian.

If you leave this out the default is a finite difference Jacobian.

So, FP=J!(FP,FS,x) or FP=J!(FP,FS,x,pdata) returns FP=F'(x).

(FP,FS, x) must be the argument list, even if FP does not need FS. One reason for this is that the finite-difference Jacobian does and that is the default in the solver.

Your Jacobian function MUST have -> return FP <- at the end. See the examples in the docstrings and in TestProblems/Systems/simple.jl

- Precision: Lemme tell ya 'bout precision. I designed this code for full precision functions and linear algebra in any precision you want. You can declare FPS as Float64, Float32, or Float16 and nsol will do the right thing if YOU do not destroy the declaration in your J! function. I'm amazed that this works so easily. If the Jacobian is reasonably well conditioned, you can cut the cost of Jacobian factorization and storage in half with no loss. For large dense Jacobians and inexpensive functions, this is a good deal.

  BUT ... There is very limited support for direct sparse solvers in anything other than Float64. I recommend that you only use Float64 with direct sparse solvers unless you really know what you're doing. I have a couple examples in the notebook, but watch out.

Keyword Arguments (kwargs):

rtol and atol: relative and absolute error tolerances

maxit: limit on nonlinear iterations

solver: default = "newton"

Your choices are "newton" or "chord". However, you have sham at your disposal only if you chose newton. "chord" will keep using the initial derivative until the iterate converges, uses the iteration budget, or the line search fails. It is not the same as sham=Inf, which is smarter.

sham: default = 5 (ie Newton)

This is the Shamanskii method. If sham=1, you have Newton. The iteration updates the derivative every sham iterations. The convergence rate has local q-order sham+1 if you only count iterations where you update the derivative. You need not provide your own derivative function to use this option. sham=Inf is chord only if chord is converging well.

I made sham=1 the default for scalar equations. For systems I'm
more aggressive and want to invest as little energy in linear
algebra as possible. So the default is sham=5.

armmax: upper bound on step size reductions in line search

resdec: default = .1

This is the target value for residual reduction. The default
value is .1. In the old MATLAB codes it was .5. I only turn
Shamanskii on if the residuals are decreasing rapidly, at least
a factor of resdec, and the line search is quiescent. If you
want to eliminate resdec from the method (you don't) then set
resdec = 1.0 and you will never hear from it again.

dx: default = 1.e-7

difference increment in finite-difference derivatives
h=dx*norm(x,Inf)+1.e-8

armfix: default = false

The default is a parabolic line search (ie false). Set to true
and the step size will be fixed at .5. Don't do this unless you
are doing experiments for research.

pdata:

precomputed data for the function/Jacobian. Things will go
better if you use this rather than hide the data in global
variables within the module for your function/Jacobian

If you use pdata in either of F! or J!, you must use it in the
calling sequence of both.

jfact: default = klfact (tries to figure out best choice)

If your Jacobian has any special structure, please set jfact to
the correct choice for a factorization.

I use jfact when I call PrepareJac! to evaluate the Jacobian
(using your J!) and factor it. The default is to use klfact
(an internal function) to do something reasonable. For general
dense matrices, klfact picks lu! to compute an LU factorization
and share storage with the Jacobian. You may change LU to
something else by, for example, setting jfact = cholesky! if
your Jacobian is spd.

klfact knows about banded matrices and picks qr. You should,
however RTFM, allocate the extra two upper bands, and use
jfact=qr! to override klfact.

klfact uses lu for general sparse matrices.

If you give me something that klfact does not know how to
dispatch on, then nothing happens. I just return the original
Jacobian matrix and nsol will use backslash to compute the
Newton step. I know that this is probably not optimal in your
situation, so it is good to pick something else, like jfact =
lu.

If you want to manage your own factorization within your
Jacobian evaluation function, then set

jfact = nofact

and nsol will not attempt to factor your Jacobian. That is
also what happens when klfact does not know what to do. Your
Jacobian is sent directly to Julia's \ operation

Please do not mess with the line that calls PrepareJac!.

    FPF = PrepareJac!(FPS, FS, x, ItRules)

FPF is not the same as FPS (the storage you allocate for the
Jacobian) for a reason. FPF and FPS do not have the same type,
even though they share storage. So, FPS=PrepareJac!(FPS, FS,
...) will break things.

printerr: default = true

I print a helpful message when the solver fails. To suppress
that message set printerr to false.

keepsolhist: default = false

Set this to true to get the history of the iteration in the
output tuple. This is on by default for scalar equations and
off for systems. Only turn it on if you have use for the data,
which can get REALLY LARGE.

stagnationok: default = false

Set this to true if you want to disable the line search and
either observe divergence or stagnation. This is only useful
for research or writing a book.

Output:

  • A named tuple (solution, functionval, history, stats, idid,
    errcode, solhist)

where

- solution = converged result

- functionval = F(solution)

- history = the vector of residual norms ($||F(x)||$) for the
iteration

- stats = named tuple of the history of (ifun, ijac, iarm), the number of functions/derivatives/steplength reductions at each iteration.

I do not count the function values for a finite-difference derivative because they count toward a Jacobian evaluation.

- idid=true if the iteration succeeded and false if not.

- errcode = 0 if the iteration succeeded

    = -1 if the initial iterate satisfies the termination criteria

    = 10 if no convergence after maxit iterations

    = 1 if the line search failed

- solhist:

This is the entire history of the iteration if you've set keepsolhist=true

solhist is an N x K array where N is the length of x and K is the number of iterations + 1. So, for scalar equations, it's a row vector.

Examples for nsol

World's easiest problem example. Test 64 and 32 bit Jacobians. No meaningful difference in the residual histories or the converged solutions.

```
julia> function f!(fv,x)
       fv[1]=x[1] + sin(x[2])
       fv[2]=cos(x[1]+x[2])
       return fv
       end
f (generic function with 1 method)

julia> x=ones(2); fv=zeros(2); jv=zeros(2,2);
julia> jv32=zeros(Float32,2,2);
julia> nout=nsol(f!,x,fv,jv; sham=1);
julia> nout32=nsol(f!,x,fv,jv32; sham=1);
julia> [nout.history nout32.history]
5×2 Matrix{Float64}:
 1.88791e+00  1.88791e+00
 2.43119e-01  2.43120e-01
 1.19231e-02  1.19231e-02
 1.03266e-05  1.03265e-05
 1.46388e-11  1.45995e-11
```

```
julia> [nout.solution nout.solution-nout32.solution]
2×2 Array{Float64,2}:
 -7.39085e-01  -5.48450e-14
  2.30988e+00  -2.26485e-14
```

H-equation example. I'm taking the sham=5 default here, so the convergence is not quadratic. The good news is that we evaluate the Jacobian only once.

```
julia> n=16; x0=ones(n); FS=ones(n); JV=ones(n,n);
julia> hdata=heqinit(x0, .5);
julia> hout=nsol(heqf!,x0,FS,JV;pdata=hdata);
julia> hout.history
4-element Array{Float64,1}:
 6.17376e-01
 3.17810e-03
 2.75227e-05
 2.35817e-07
```

The calling sequence for the Newton solvers in this book are similar, differing mostly in the management of the linear solver and memory allocation. The calling sequence for **nsol.jl** is

```
function nsol(
    F!,
    x0,
    FS,
    FPS,
    J! = diffjac!;
    rtol = 1.e-6,
    atol = 1.e-12,
    maxit = 20,
    solver = "newton",
    sham = 1,
    armmax = 10,
    resdec = 0.1,
    dx = 1.e-7,
    armfix = false,
    pdata = nothing,
    jfact = lu!,
    printerr = true,
    keepsolhist = false,
    stagnationok = false,
)
```

As we said earlier in the chapter, the calling sequence has some new things which are not in **nsolsc.jl**. The most significant are the arrays **FS** and **FPS**, which preallocate storage for the function and Jacobian. As we have pointed out earlier, the farther upstream one allocates memory the better, so **nsol.jl** insists that you allocate a vector **FS** of the same size as the initial iterate and a matrix **FPS** for the Jacobian.

For problems in several variables, the keyword argument **pdata** is very important. This is the data structure for you to store any precomputed or preallocated data your function evaluation needs. You will almost surely need **pdata** for any but the most trivial problems. Most of the examples from in this chapter use **pdata** in a serious manner.

You should dimension **x0** either as (N) (or equivalently $(N,)$) and dimension **FS** the same way. **nsol.jl** expects vectors to be in double precision (Float64). Unless you are porting old code you should dimension **x** and **F** as vectors (N), not as two-dimensional arrays with one column $(N, 1)$. The solvers should handle both, but I cannot guarantee that.

The **!** in the function evaluation **F!** is to indicate that **nsol.jl** expects **F** to overwrite its input. So, the way to call **F!** is to preallocate the storage for the function value in an array **FS** and then call the function as

```
F!(FS,x)
```

or

```
F!(FS,x,pdata)
```

nsol.jl will figure out if you have populated **pdata** or left it alone as the default value of nothing.

And now for the Jacobian. **nsol.jl** uses direct methods for linear algebra. If your matrix is dense, the default is to use Julia's **lu!** function to do an LU factorization. If your matrix is symmetric or symmetric positive definite, you can use the **jfact** keyword to change **lu!** to **ldlt!**, or **cholesky!**, for example. **nsol.jl** assumes that the factorization you ask for will overwrite the matrix. Hence, the **factorize** function in Julia is not what you want for this application.

If the Jacobian is sparse, **nsol.jl** will use **lu** instead of **lu!**. The reason for this is that one cannot overwrite the Jacobian with the factorization in the general sparse case.

You will also need to preallocate storage for the Jacobian in the array **FPS**. You may use any legal real precision for **FPS**. Float64 is the default. If you use Float32, you cut the storage for the matrix and the time for the factorization in half. We recommend that you do this if your Jacobian is dense. If you are using the **SuiteSparse.jl** sparse solvers, then you must store the Jacobian in double precision. **SuiteSparse.jl** does not support lower precision.

Your Jacobian computation **J!** must also overwrite its input. The call looks like

```
J!(FS,FP,x)
```

or

```
J!(FS,FP,x,pdata)
```

returns FP=F'(x). The input FP=F(x), which **nsol.jl** has already computed, has to be there.

2.8.2 ▪ H-Equation Revisited

In this section we do several experiments to illustrate the advantages of infrequent evaluation and factorization of the Jacobian and mixed precision computation. We will begin

with a function that will support our testing. We want to investigate combinations of Newton's method/Shamanskii with sham=5 (the default), storing and factoring the Jacobian in
double/single precision, and analytic/forward difference Jacobians. To make this easy we
write a function that solves the H-equation with **nsol.jl** and lets us vary these cases.

The functions for the residual **heqf!.jl**, the Jacobian **heqJ!.jl**, and the precomputed data
heqinit.jl are in the large file **Hequation.jl**.

I'm passing the precomputed data to the function rather than computing it within. This
keeps the cost of the precomputed data out of the benchmarking I'll do later.

We use **splat** in this example. We populate a named tuple bargs to keep the keyword
arguments in a convenient place and then, when it's time to give it to **nsol.jl**, the call looks
like bargs.... The three dots are the **splat** and tell **nsol** to expand bargs and harvest the
keyword arguments.

```
[3]: function htest(x0, FS, FPS, hdata; analytic=false, hsham=5)
         n=length(FS)
         #
         # I've preallocaed x0, FS, and FPS. But they may have
         # been changed by previous runs.
         # The cost of resetting their entries to 1.0 is insignificant.
         #
         FS.=1.0
         FPS.=1.0
         bargs=(atol = 1.e-10, rtol = 1.e-10, sham = hsham,
             resdec = .1, pdata=hdata)
         if analytic
             nout=nsol( heqf!, x0, FS, FPS, heqJ!; bargs...)
         else
             nout=nsol( heqf!, x0, FS, FPS; bargs...)
         end
         return nout
     end
```

```
[3]: htest (generic function with 1 method)
```

To begin we will compare the iteration histories for four cases. We will consider analytic
and forward difference Jacobians with the storage and factorization of the Jacobian done in
double and single precision. **Theorem 1.2** says the results should be almost indistinguishable. We will begin with Newton's method.

All we need to do to store and factor Jacobians is to allocate the storage in single precision.
That allocation is the line FPS32=ones(Float32,n,n);. Note that we must reset FS and
FPS before each new call to **nsol.jl** because the solver uses the storage for residuals and
Jacobians for the entire iteration. We do this with **broadcast** after the initial allocation
.=1.0 instead of =ones(n,n) to avoid reallocation of the Jacobian.

We will print all the residual histories in an array. The history vectors are the same length
and are very hard to tell apart until the residual norm is one iteration from stagnation. This
is just what the theory predicts. The theory (see [113]) also predicts that there will be little

difference between double precision linear algebra and single precision. We observe this as well.

```
[4]:  n=1024; FS=ones(n); FPS=ones(n,n); FPS32=ones(Float32,n,n);
      x0=ones(n); c=.5; hdata = heqinit(x0, c);
      nouta64=htest(x0, FS, FPS, hdata;
          analytic=true, hsham=1);
      nouta32=htest(x0, FS, FPS32, hdata;
          analytic=true, hsham=1);
      noutfd64=htest(x0, FS, FPS, hdata;
          analytic=true, hsham=1);
      noutfd32=htest(x0, FS, FPS32,hdata;
          analytic=false, hsham=1);
      hinit=nouta64.history[1]
      ha64=nouta64.history./hinit
      ha32=nouta32.history./hinit
      hd64=noutfd64.history./hinit
      hd32=noutfd32.history./hinit
      [ha64 ha32 hd64 hd32]
```

```
[4]:  4×4 Matrix{Float64}:
       1.00000e+00   1.00000e+00   1.00000e+00   1.00000e+00
       5.14148e-03   5.14131e-03   5.14148e-03   5.14112e-03
       1.00479e-07   1.00385e-07   1.00479e-07   1.04305e-07
       2.12995e-15   8.11565e-14   2.12995e-15   1.97470e-13
```

We will now do the same thing with the default setting of sham=5. The theory correctly predicts that we will see more nonlinear iterations. We will be using **BenchmarkTools.jl** to compare the costs later.

```
[5]:  nouta64=htest(x0, FS, FPS, hdata; analytic=true);
      noutfd64=htest(x0, FS, FPS, hdata; analytic=false);
      nouta32=htest(x0, FS, FPS32, hdata; analytic=true);
      noutfd32=htest(x0, FS, FPS32, hdata; analytic=false);
      hinit=nouta64.history[1]
      ha64=nouta64.history./hinit
      ha32=nouta32.history./hinit
      hd64=noutfd64.history./hinit
      hd32=noutfd32.history./hinit
      [ha64 ha32 hd64 hd32]
```

```
[5]:  6×4 Matrix{Float64}:
       1.00000e+00   1.00000e+00   1.00000e+00   1.00000e+00
       5.14148e-03   5.14131e-03   5.14126e-03   5.14112e-03
       4.44954e-05   4.44938e-05   4.44920e-05   4.44910e-05
       3.81019e-07   3.81006e-07   3.80979e-07   3.80968e-07
       3.26071e-09   3.26058e-09   3.26027e-09   3.26014e-09
       2.79020e-11   2.79000e-11   2.78989e-11   2.78966e-11
```

This is interesting. We need more iterations, but we evaluate the Jacobian only once for Shamanskii. We can see this by looking at the `stats` field of the output tuple. They are all the same, so we will use `nouta64`.

```
[6]: nouta64.stats
```

```
[6]: (ifun = [1, 1, 1, 1, 1, 1], ijac = [0, 1, 0, 0, 0, 0],
      iarm = [0, 0, 0, 0, 0, 0])
```

The interpretation is that we do a function evaluation at all iterations and a single Jacobian evaluation to compute x_1. We do no Jacobian work after that. The default in **nsol.jl** is to reevaluate the Jacobian if the reduction in the residual norm larger than `resdec = .1`. You can change `resdec` in the keyword arguments. The `iarm` field in the stats tells us that a line search was not necessary.

We will use the **BenchmarkTools.jl** package to look at performance. The `@btime` command will show compute time and memory allocations for an average of several runs. The averaging will mitigate the effects of the compile time for the first run.

To begin, we will compare the four versions of Newton's method. Note the $ in front of the array arguments. This is **interpolation** and using it is important if you want to get accurate results from `@btime`.

```
[7]: println("analytic, double");
     @btime htest($x0, $FS, $FPS, hdata;
         analytic=true, hsham=1);
     println("finite difference, double");
     @btime htest($x0, $FS, $FPS, hdata;
         analytic=false, hsham=1);
     println("analytic, single");
     @btime htest($x0, $FS, $FPS32, hdata;
         analytic=true, hsham=1);
     println("finite difference, single");
     @btime htest($x0, $FS, $FPS32, hdata;
         analytic=false, hsham=1);
```

```
analytic, double
  13.502 ms (6189 allocations: 468.00 KiB)
finite difference, double
  75.275 ms (9264 allocations: 731.95 KiB)
analytic, single
  6.541 ms (6195 allocations: 492.75 KiB)
finite difference, single
  70.256 ms (9270 allocations: 756.70 KiB)
```

Even for a problem of only modest size, the differences between the analytic Jacobian and the forward difference are significant. The differences between single and double precision linear algebra are, at least for the analytic Jacobian, roughly the factor of two we'd expect if the matrix factorization dominated the computation. For the forward difference Jacobian, we see that the cost of the Jacobian evaluation dominates everything else.

Next, we look at the default sham=5 from **nsol.jl**.

```
[8]: println("analytic, double");
     @btime htest($x0, $FS, $FPS, hdata; analytic=true);
     println("finite difference, double");
     @btime htest($x0, $FS, $FPS, hdata; analytic=false);
     println("analytic, single");
     @btime htest($x0, $FS, $FPS32, hdata; analytic=true);
     println("finite difference, single");
     @btime htest($x0, $FS, $FPS32, hdata; analytic=false);
```

```
analytic, double
  4.666 ms (2091 allocations: 243.62 KiB)
finite difference, double
  25.552 ms (3116 allocations: 331.61 KiB)
analytic, single
  2.639 ms (2101 allocations: 284.88 KiB)
finite difference, single
  23.881 ms (3126 allocations: 372.86 KiB)
```

The moral here is pretty clear. We see that compute time is cut by a factor of at least two over Newton's method in all cases. The result for an analytic Jacobian with linear algebra in single precision and sham=5 is 19 times faster than our slowest computation (Newton + finite difference Jacobian + double precision linear algebra). So, do less linear algebra and do it in single precision.

Finally, we will increase the dimension. As we do that the computation becomes more burdensome, so we will only do four cases, all with an analytic Jacobian. The size of this example is large enough to clearly show the factor of two reduction in cost one gets from a single precision Jacobian.

The reader with lots of free time should try these cases with a forward difference Jacobian.

```
[9]: n=4096; FS=ones(n); FPS=ones(n,n);
     FPS32=ones(Float32,n,n);
     x0=ones(n); c=.5; hdata = heqinit(x0, c);
     println("analytic, double, Newton");
     @btime htest($x0, $FS, $FPS, hdata;
         analytic=true, hsham=1);
     println("analytic, single, Newton");
     @btime htest($x0, $FS, $FPS32, hdata;
         analytic=true, hsham=1);
     println("analytic, double, sham=5");
     @btime htest($x0, $FS, $FPS, hdata;
         analytic=true);
     println("analytic, single, sham=5");
     @btime htest($x0, $FS, $FPS32, hdata;
         analytic=true);
```

```
analytic, double, Newton
```

```
  405.242 ms (24643 allocations: 1.81 MiB)
analytic, single, Newton
  196.254 ms (24649 allocations: 1.91 MiB)
analytic, double, sham=5
  161.711 ms (8253 allocations: 962.22 KiB)
analytic, single, sham=5
  77.529 ms (8263 allocations: 1.10 MiB)
```

For sufficiently large dimension, the linear algebra cost will dominate the computation. Time should increase by roughly a factor of 8 as the dimension doubles because our LU factorization takes $O(N^3)$ operations. We see that with this computation. Increasing the dimension from 1024 to 4096 did increase the runtimes by (roughly) a factor of 64, and we see that using single precision Jacobians cuts the runtime in half. Note also that the single precision Shamanskii run is now five times faster than the double precision Newton computation. One of the projects at the end of this chapter challenges you to increase the dimension and compare the timings as you do that. Remember that we allocated storage for the Jacobian when we defined FPS, so @btime is not measuring the allocation for the Jacobian.

2.8.3 ▪ More on the Two-Point BVP

If the Jacobian is sparse and you use the solvers from **SuiteSparse.jl**, then you cannot use single precision. Even if the structure of the Jacobian allows you to use the LAPACK solvers or a special-purpose package, there is less benefit in using single precision for linear algebra than in the dense case. We will explore that for the boundary value problem, where we can use **BandedMatrices.jl** [144] and **qr!** for the linear solver. **qr!** supports single precision linear algebra, so we will use that. **lu!** does not, so the support is not consistent.

Then we will set up the problem for a very fine mesh, far finer than one needs to get a useful result, to illustrate the performance. The band solver takes $O(N)$ work, so we would expect the solve to be fast. To set things up we mimic **bvp_solve.jl** .

[10]:
```
# Set it up
    n=10^5;
    bdata = bvpinit(n, Float64);
#
    U0 = zeros(2n);
    FS = zeros(2n);
# Banded matrix with the correct number of bands
# Make double and single precision copies
    FPV = BandedMatrix{Float64}(Zeros(2n, 2n), (2, 4));
    FPV32 = BandedMatrix{Float32}(Zeros(2n, 2n), (2, 4));
#
# Build the initial iterate
#
    tv = bdata.tv;
    sv = -.1 * tv .* tv;
    view(U0,1:2:2n-1) .= exp.(-.1 .* tv .* tv);
    view(U0,2:2:2n).= -.2 .* view(U0,1:2:2n-1) .* tv;
```

We will begin by comparing the default sham=5 with Newton's method. As we did with the H-equation, we will write a test function that uses the data we allocated above. We only use an analytic Jacobian for this and other examples with sparse Jacobians. The reader might want to look at the project in this chapter on sparse differencing in section 2.9.4. We will look at the convergence in Figure 2.8

```
[11]:  function bvptest(U0, FS, FPS, bdata; bsham=5, bfact=qr!)
           FS .*= 0.0
           FPS .*= 0.
               bvpout = nsol(Fbvp!, U0, FS, FPS, Jbvp!;
               atol=1.e-8, rtol = 1.e-8, sham=bsham,
                   pdata = bdata, jfact=bfact)
           return bvpout
       end
```

First we will look at the convergence by running four cases with single and double precision and the kwarg sham set to 1 and 5.

```
[12]:  bvpoutn64=bvptest(U0, FS, FPV, bdata;  bsham=1);
       bvpouts64=bvptest(U0, FS, FPV, bdata;  bsham=5);
       bvpoutn32=bvptest(U0, FS, FPV32, bdata;  bsham=1);
       bvpouts32=bvptest(U0, FS, FPV32, bdata;  bsham=5);
       newtonhist=bvpoutn64.history./bvpoutn64.history[1];
       shamhist=bvpouts64.history./bvpoutn64.history[1];
       newtonhist32=bvpoutn32.history./bvpoutn64.history[1];
       shamhist32=bvpouts32.history./bvpoutn64.history[1];
       nn=length(newtonhist);
       ns=length(shamhist);
       nn32=length(newtonhist32);
       ns32=length(shamhist32);
       semilogy(0:nn-1,newtonhist,"k-",0:ns-1,shamhist,"k--",
           0:nn32-1,newtonhist32,"k.",0:ns32-1,shamhist32,"k-.")
       legend(["newton","sham=5","newton, single",
               "sham=5, single"]);
```

We now query the iteration statistics to see what the line search did for Newton's method in double precision. We see that the line search was active in the middle of the iteration.

```
[13]:  bvpoutn64.stats.iarm'
```

```
[13]:  1×10 adjoint(::Vector{Int64}) with eltype Int64:
        0  0  0  2  2  1  0  0  0  0
```

Updating the Jacobian less frequently has a smaller impact than in the case of the H-equation. You should expect this because the linear solver costs $O(N)$ work rather than $O(N^3)$ and the Jacobian is updated for the early iterations because the residual is not decreasing fast enough.

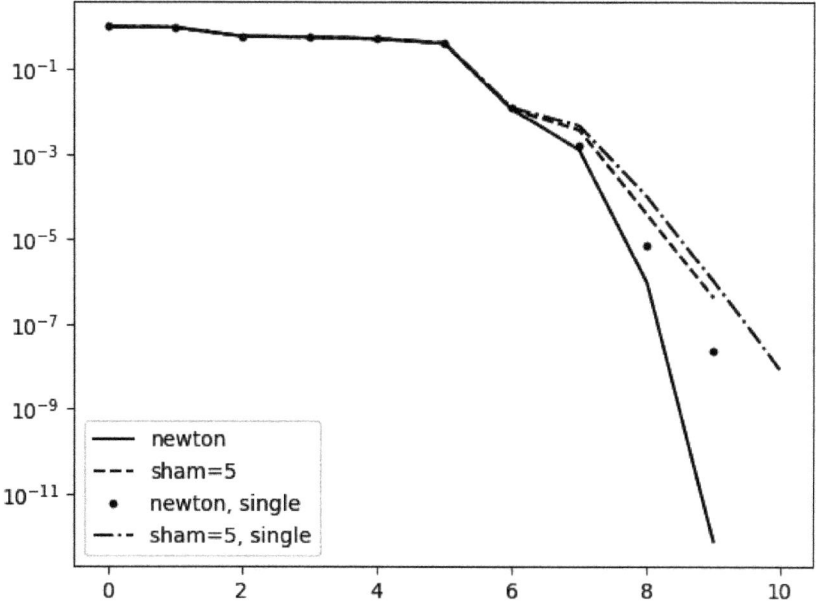

Figure 2.8: Residual histories for the BVP.

The other interesting feature of this computation is that the single precision results differ in a meaningful way from the double precision results. The reason for this is that the Jacobian is ill-conditioned enough to affect the quality of the single precision solver. The condition number of the Jacobian is $O(N)$, so for $N = 10^5$ one should expect different results from the linear solve in single precision, and that's what you get.

Moreover, the benchmark results say that there is no benefit from doing the linear algebra in single precision. This is also no surprise since the factorization is $O(N)$ work.

The reader should try this problem with $N = 10^6$ and watch the line search fail for the single precision linear algebra computations. That is also no surprise with an inaccurate Newton direction.

```
[14]: println("analytic, double, Newton");
      @btime bvptest($U0, $FS, $FPV, $bdata;  bsham=1);
      println("analytic, double, sham=5");
      @btime bvptest($U0, $FS, $FPV, $bdata;  bsham=5);
      println("analytic, single, Newton");
      @btime bvptest($U0, $FS, $FPV32, $bdata;  bsham=1);
      println("analytic, single, sham=5");
      @btime bvptest($U0, $FS, $FPV32, $bdata;  bsham=5);
```

```
analytic, double, Newton
  152.681 ms (93 allocations: 47.31 MiB)
analytic, double, sham=5
  127.656 ms (87 allocations: 44.25 MiB)
```

```
analytic, single, Newton
  143.321 ms (129 allocations: 54.17 MiB)
analytic, single, sham=5
  122.475 ms (131 allocations: 57.23 MiB)
```

2.8.4 ▪ Shamanskii for the Convection-Diffusion Problem

The dominant cost in the solution of the PDE problem is the computation and factorization of the sparse Jacobian. Even for the small problem with $31^2 = 961$ unknowns, one can see the effects. We will compare a Newton iteration with the default strategy using @btime. The code that generated the results in section 2.7.5 contained the initialization. It's important to separate that from the solver if you're doing benchmarking. So we will setup the problem first and then benchmark the solver phase.

[15]:
```julia
n=31;
# Get some room for the residual
u0=zeros(n*n);
FS=copy(u0);
# Get the precomputed data from pdeinit
pdata=pdeinit(n)
# Storage for the Jacobian,
# same sparsity pattern as the discrete Laplacian
J=copy(pdata.D2);
println("Newton");
@btime nsol(pdeF!, u0, FS, J, pdeJ!;
    resdec=.5, rtol=1.e-7, atol=1.e-10,
    pdata=pdata, sham=1);
println("sham=5");
@btime nsol(pdeF!, u0, FS, J, pdeJ!;
    resdec=.5, rtol=1.e-7, atol=1.e-10,
    pdata=pdata, sham=5);
```

```
Newton
  9.232 ms (539 allocations: 9.51 MiB)
sham=5
  6.329 ms (382 allocations: 6.55 MiB)
```

The Shamanskii iteration, which forms and factors the Jacobian only twice, saves about 40% in time. This is a larger savings than we saw with the two-point boundary value problem, but far from dramatic. The **SuiteSparse** solvers are very good. The reader should try the same test with a finer mesh to see if the results change.

2.8.5 ▪ ptcsol.jl

ptcsol.jl is our (ΨTC) solve. As usual, we begin with the docstrings.

[16]:
```julia
?ptcsol
```

[16]:
```
ptcsol(F!, x0, FS, FPS, J! = diffjac!; rtol=1.e-6, atol=1.e-12,
maxit=20, delta0=1.e-6, dx=1.e-7, pdata = nothing, jfact =
klfact, printerr = true, keepsolhist = false, jknowsdt = false)
```

C. T. Kelley, 2022

Julia versions of the nonlinear solvers from my SIAM books.
Herewith: some new stuff ==> ptcsol

PTC finds the steady-state solution of u' = -F(u), u(0) = u_0.
The - sign is a convention.

You must allocate storage for the function and Jacobian in
advance -> in the calling program <- ie. in FS and FPS

Inputs:

- F!: function evaluation, the ! indicates that F!
  overwrites FS, your preallocated storage for the function.

  So, FS=F!(FS,x) or FS=F!(FS,x,pdata) returns FS=F(x)

  Your function MUST have -> return FS <- at the end. See the
  examples in the TestProblems/Systems/FBeam!.jl

- x0: initial iterate

- FS: Preallocated storage for function. It is a vector of
  size N

  You should store it as (N) and design F! to use vectors of
  size (N). If you use (N,1) consistently instead, the solvers
  may work, but I make no guarantees.

- FPS: preallocated storage for Jacobian. It is an N x N
  matrix

  If FPS is sparse, you **must** allocate storage for the diagonal
  so I will have room to put 1/dt in there.

- J!: Jacobian evaluation, the ! indicates that J!
  overwrites FPS, your preallocated storage for the Jacobian.
  If you leave this out the default is a finite difference
  Jacobian.

  So, FP=J!(FP,FS,x) or FP=J!(FP,FS,x,pdata) returns FP=F'(x);
  (FP,FS, x) must be the argument list, even if FP does not
  need FS. One reason for this is that the finite-difference
  Jacobian does and that is the default in the solver.

  Your Jacobian function MUST have -> return FP <- at the end.
  See the examples in the TestProblems/Systems/FBeam!.jl

  You may have a better way to add (1/dt) I to your Jacobian.
  If you want to do this yourself then your Jacobian function

should be `FP=J!(FP,FS,x,dt)` or `FP=J!(FP,FS,x,dt,pdata)` and return `F'(x) + (1.0/dt)*I`.

You will also have to set the kwarg **jknowsdt** to true.

- Precision: Lemme tell ya 'bout precision. I designed this code for full precision functions and linear algebra in any precision you want. You can declare FPS as Float64, Float32, or Float16 and ptcsol will do the right thing if YOU do not destroy the declaration in your J! function. I'm amazed that this works so easily. If the Jacobian is reasonably well conditioned, you can cut the cost of Jacobian factorization and storage in half with no loss. For large dense Jacobians and inexpensive functions, this is a good deal.

  BUT ... There is very limited support for direct sparse solvers in anything other than Float64. I recommend that you only use Float64 with direct sparse solvers unless you really know what you're doing. I have a couple examples in the notebook, but watch out.

Keyword Arguments (kwargs):

rtol and atol: relative and absolute error tolerances

delta0: initial pseudo time step. The default value of 1.e-3 is a bit conservative and is one option you really should play with. Look at the example where I set it to 1.0!

maxit: limit on nonlinear iterations, default=100.

This is coupled to delta0. If your choice of delta0 is too small (conservative) then you'll need many iterations to converge and will need a larger value of maxit

For PTC you'll need more iterations than for a straight-up nonlinear solve. This is part of the price for finding the stable solution.

dx: default = 1.e-7

difference increment in finite-difference derivatives h=dx*norm(x)+1.e-6

pdata:

precomputed data for the function/Jacobian. Things will go better if you use this rather than hide the data in global variables within the module for your function/Jacobian

jfact: default = klfact (tries to figure out best choice)

If your Jacobian has any special structure, please set jfact to the correct choice for a factorization.

I use jfact when I call PTCUpdate to evaluate the Jacobian (using your J!) and factor it. The default is to use klfact (an internal function) to do something reasonable. For general dense matrices, klfact picks lu! to compute an LU factorization and share storage with the Jacobian. You may change LU to something else by, for example, setting jfact = cholesky! if your Jacobian is spd.

klfact knows about banded matrices and picks qr. You should, however RTFM, allocate the extra two upper bands, and use jfact=qr! to override klfact.

klfact uses lu for general sparse matrices.

If you give me something that klfact does not know how to dispatch on, then nothing happens. I just return the original Jacobian matrix and ptcsol will use backslash to compute the Newton step.

I know that this is probably not optimal in your situation, so it is good to pick something else, like jfact = lu.

printerr: default = true

I print a helpful message when the solver fails. To suppress that message set printerr to false.

keepsolhist: default = false

Set this to true to get the history of the iteration in the output tuple. This is on by default for scalar equations and off for systems. Only turn it on if you have use for the data, which can get REALLY LARGE.

jknowsdt: default = false

Set this to true if your Jacobian evaluation function returns $F'(x) + (1/dt) I$. You'll also need to follow the rules above for the Jacobian evaluation function. I do not recommend this and if your Jacobian is anything other than a matrix I can't promise anything. I've tested this for matrix outputs only.

Output:

A named tuple (solution, functionval, history, stats, idid, errcode, solhist) where

solution = converged result functionval = F(solution) history = the vector of residual norms ($||F(x)||$) for the iteration

Unlike nsol, nsoli, or even ptcsoli, ptcsol has a fixed cost per iteration of one function, one Jacobian, and one Factorization. Hence iteration statistics are not interesting and not in the output.

idid=true if the iteration succeeded and false if not.

errcode = 0 if the iteration succeeded = -1 if the initial iterate satisfies the termination criteria = 10 if no convergence after maxit iterations

solhist:

This is the entire history of the iteration if you've set keepsolhist=true

solhist is an N x K array where N is the length of x and K is the number of iteration + 1. So, for scalar equations, it's a row vector.

Example for ptcsol

The buckling beam problem. You'll need to use TestProblems for this to work.

```
julia> using SIAMFANLEquations.TestProblems

julia> n=63; maxit=1000; delta = 0.01; lambda = 20.0;

julia> bdata = beaminit(n, 0.0, lambda); x = bdata.x;

julia> u0 = x .* (1.0 .- x) .* (2.0 .- x);

julia> u0 .*= exp.(-10.0 * u0);

julia> FS = copy(u0); FPS = copy(bdata.D2);

julia> pout = ptcsol( FBeam!, u0, FS, FPS, BeamJ!;
 rtol = 1.e-10, pdata = bdata, delta0 = delta, maxit = maxit);

julia> # It takes a few iterations to get there.
       length(pout.history)
25

julia> [pout.history[1:5] pout.history[21:25]]
5x2 Array{Float64,2}:
 6.31230e+01  9.75412e-01
 7.52624e+00  8.35295e-02
 8.31545e+00  6.58797e-04
 3.15455e+01  4.12697e-08
```

```
   3.66566e+01   6.29295e-12

julia> # We get the nonnegative steady state.
       maximum(pout.solution)
2.19086e+00
```

2.8.6 ▪ More on the Buckling Beam

The residual function for the beam uses precomputed data to store the discrete second derivative and the bifurcation parameter λ. These data are smaller and far less complex than the data for the H-equation, and it's worthwhile to look at the functions. These are part of the **TestProblems** submodule for **SIAMFANLEquations.jl**. We will not cover all the details here, but will give you enough to see how precomputed data is useful. The file is **TestProblems/Systems/FBeam!.jl**.

First the function itself:

```
"""
FBeam!(FS, U, bdata)
Function evaluation for PTC example.
F(u) = -u'' - lambda sin(u)
"""
function FBeam!(FS, U, bdata)
    D2 = bdata.D2
    lambda = bdata.lambda
    su = lambda * sin.(U)
    FS .= (D2 * U - su)
end
```

Note how the function FBeam harvests λ and the discrete second derivative D2 from the precomputed data bdata. In this case bdata is a **named tuple** which we create with the **beaminit** function:

```
"""
beaminit(n,dt,lambda=20.0)

Set up the beam problem with n interior grid points.
"""
function beaminit(n, dt, lambda = 20.0)
    D2 = Lap1d(n)
    dx = 1.0 / (n + 1)
    x = collect(dx:dx:1.0-dx)
    UN = zeros(size(x))
    bdata = (D2 = D2, x = x, dx = dx, dt = dt,
             lambda = lambda, UN = UN)
end
```

The bdata structure has more than just D2 and λ. There are also data for the solver and the construction of the Jacobian.

Finally, the function **Lap1D** computes D2 as a tridiagonal matrix. This is part of the LinearAlgebra package that is part of Julia.Base.

```
"""
Lap1d(n)

returns -d^2/dx^2 on [0,1] zero BC
"""
function Lap1d(n)
    dx = 1 / (n + 1)
    d = 2.0 * ones(n)
    sup = -ones(n - 1)
    slo = -ones(n - 1)
    D2 = Tridiagonal(slo, d, sup)
    D2 = D2 / (dx * dx)
    return D2
end
```

2.9 ▪ Projects

2.9.1 ▪ More on the H-Equation

- Run the H-equation tests with $c = 0.99$ and larger dimensions (2^k for $k = 12, 13, 14$). Are the values of sham and resdec optimal for this case? How, for example, is resdec = Inf (your author's favorite value). To see why I like sham=Inf have a look at [21].

- What happens with $c = 1$? Is the chord method a good idea? Don't look at [113], [48], or [49] before trying to figure things out on your own.

- Increase the dimension as much as you can and compare single and double precision linear algebra.

- Duplicate the table on page 125 of [32].

2.9.2 ▪ Automatic Differentiation

(**Not trivial!**) Try using **automatic differentiation** to compute the Jacobian for the examples in this chapter and use the results to write a Jacobian evaluation function to pass to **nsol.jl**. You will first have to decide how you want to do that. Two mature packages are **ForwardDiff.jl** [164] and **Zygote.jl** [98]. Books like [80] explain the algorithmic differences between these packages.

How does the performance compare to an analytic or forward difference Jacobian?

2.9.3 ▪ Nested Iteration

Solving a differential or integral equation by **nested iteration** or **grid sequencing** means resolving the rough features of the solution of a differential or integral equation on a coarse mesh, interpolating the solution to a finer mesh, resolving on the finer mesh, and then repeating the process until you have a solution on a target mesh.

Apply this idea to some of the examples in the text, using piecewise linear interpolation to move from coarse to fine meshes. If the discretization is second-order accurate and you halve the mesh width at each level, how should you terminate the solver at each level? What kind of iteration statistics would tell you that you've done a satisfactory job?

2.9.4 ▪ Sparse Differencing

Write Jacobian evaluation functions using sparse differencing for the boundary value problem and the buckling beam examples. You can try, for example, the methods from [41] or [37]. [111] has a MATLAB code for banded matrices that, while convertible to Julia, needs some work to explicitly store the Jacobian as a BandedMatrix so BandedMatrices.jl will solve the linear system efficiently. Another alternative is to use **sparsefdiff** [160] from **SparseDiffTools.jl**.

No matter what you do, you'll need to think about fill-in, symbolic factorization in the general case, and storage.

Have fun!

2.9.5 ▪ lu! for Sparse Matrices

This project is for Julia 1.5 and higher. The option `jfact=nofact` in **nsol.jl** is there so you can build and manage your own factorization inside your Jacobian evaluation function. Use this feature to store a factorization of the Jacobian for the PDE problem and reuse that storage rather than reallocating with every call to lu. What are the savings in time and allocations? You'll need to figure out how to use `lu!` for sparse matrices to do this project.

2.9.6 ▪ How Does δ_0 Depend on δ_x?

Vary δ_x and δ_0 in the buckling beam example. For the fixed $\delta_x = 1/64$ in the example, explore the effects of increasing δ_0. Can you converge to the "wrong" (i.e., negative) solution in this way? What happens if you fix δ_0 and reduce δ_x?

Chapter 3

Newton–Krylov Methods

Files for This Chapter

- From the Package repository:
    - Solvers using Newton–Krylov methods: **/src/Solvers**
        * Newton's method: **nsoli.jl**
        * Pseudo-transient continuation: **ptcsoli.jl**
    - Krylov linear solvers: **/src/Solvers/LinearSolvers**
        * GMRES: **kl_gmres.jl**
        * BiCGSTAB: **kl_bicgstab.jl**
    - Test problems: **/src/TestProblems/Systems**
        * H-equation: **Hequation.jl**
        * Convection-diffusion equation: **EllipticPDE.jl** and **PDE_Tools.jl**
        * Buckling beam: **FBeam!.jl**
- From the Notebook repository: **/src/Chapter3**
  Julia codes that generate the figures and tables

3.1 ▪ Newton-Iterative Methods

Recall from section 1.4 that an inexact Newton method approximates the Newton direction with a vector $\mathbf{d}$ such that

$$\|\mathbf{F}'(\mathbf{x}_n)\mathbf{d} + \mathbf{F}(\mathbf{x}_n)\| \leq \eta \|\mathbf{F}(\mathbf{x}_n)\|. \tag{3.1}$$

The parameter η is called the **forcing term**.

Newton iterative methods realize the inexact Newton condition (3.1) by applying a linear iterative method to the equation for the Newton step and terminating that iteration when

95

(3.1) holds. We sometimes refer to this linear iteration as an **inner iteration**. Similarly, the nonlinear iteration (the while loop in Algorithm **nsol**) is often called the **outer iteration**.

The Newton–Krylov methods, as the name suggests, use Krylov subspace-based linear solvers. The methods differ in storage requirements, cost in evaluations of **F**, and robustness. Our solver package **SIAMFANLEquations.jl** has two Newton–Krylov nonlinear solvers, **src/Solvers/nsoli.jl** and **src/Solvers/ptcsoli.jl**. We include two Krylov linear solvers, GMRES [170] in the function **kl_gmres.jl** and BiCGSTAB [197] in the function **kl_bicgstab.jl**. Our linear solvers have been designed to work well with nonlinear solvers. For example, information from the nonlinear solver can be used to adjust the forcing term in the linear solver. Following convention, we will refer to the nonlinear methods as Newton-GMRES and Newton-BiCGSTAB.

3.2 ▪ Krylov Methods for Solving Linear Equations

Krylov iterative methods approximate the solution of a linear system $\mathbf{Ax} = \mathbf{b}$ with a sum of the form

$$\mathbf{x}_k = \mathbf{x}_0 + \sum_{j=0}^{k-1} \gamma_k \mathbf{A}^k \mathbf{r}_0,$$

where $\mathbf{r}_0 = \mathbf{b} - \mathbf{Ax}_0$ and $\mathbf{x}_0$ is the initial iterate. If the goal is to approximate a Newton step, as it is here, the most sensible initial iterate is $\mathbf{x}_0 = 0$, because we have no a priori knowledge of the direction, but, at least in the local phase of the iteration, expect the norm of the step to be small.

We express this in compact form as $\mathbf{x}_k \in \mathcal{K}_k$, where the kth **Krylov subspace** is

$$\mathcal{K}_k = \text{span}(\mathbf{r}_0, \mathbf{Ar}_0, \ldots, \mathbf{A}^{k-1}\mathbf{r}_0).$$

Krylov methods build the iteration by evaluating matrix-vector products, in very different ways, to build an iterate in the appropriate Krylov subspace. Our nonlinear solvers, like most implementations of Newton–Krylov methods, approximate Jacobian-vector products with forward differences as the default (see section 3.4.1). You have the option of providing an analytic Jacobian-vector project. Unlike the case for direct linear solvers, there is rarely any gain in performance or robustness in using an analytic Jacobian-vector product. If you find that the iteration is stagnating, your first step should be to reduce the forcing term. After that you might see if an analytic Jacobian-vector product helps, but it is not likely that it will.

3.2.1 ▪ GMRES

The easiest Krylov method to understand is the GMRES [170] method, the default linear solver in **nsoli.jl**. The kth GMRES iterate is the solution of the linear least squares problem of minimizing

$$\|\mathbf{b} - \mathbf{Ax}_k\|^2$$

over $\mathcal{K}_k$. We refer the reader to [53, 107, 170, 195] for the details of the implementation, pointing out only that it is not a trivial task to implement GMRES well.

GMRES must accumulate the history of the linear iteration as an orthonormal basis for the Krylov subspace. This is an important property of the method because one can, and often does for large problems, exhaust the available fast memory. Any implementation of

GMRES must limit the size of the Krylov subspace. GMRES(m) does this by restarting the iteration when the size of the Krylov space exceeds m vectors. With **nsoli.jl** you must allocate storage of $m + 1$ vectors for the Krylov basis for GMRES(m). Then if you ask for more than m GMRES iterations the iteration will restart. The default is to take m GMRES iterations or fewer before termination. The convergence theory for GMRES does not apply to GMRES(m) and the performance of GMRES(m) can be poor if m is small.

GMRES, like other Krylov methods, is often, but not always, implemented as a **matrix-free** method. The reason for this is that only matrix-vector products, rather than details of the matrix itself, are needed to implement a Krylov method. One example of a matrix-free method is the use of a forward difference Jacobian-vector product (see section 3.4.1).

Convergence of GMRES

As a general rule (but not an absolute law! [142]), GMRES, like other Krylov methods, performs best if the eigenvalues of $\mathbf{A}$ are in a few tight clusters [28, 53, 107, 195]. One way to understand this, keeping in mind that $\mathbf{x}_0 = 0$, is to observe that the kth GMRES residual is in $\mathcal{K}_k$ and hence can be written as a polynomial in $\mathbf{A}$ applied to the residual

$$\mathbf{r}_k = \mathbf{b} - \mathbf{A}\mathbf{x}_k = p(\mathbf{A})\mathbf{r}_0 = p(\mathbf{A})\mathbf{b}.$$

Here $p \in \mathcal{P}_k$, the set of kth-degree **residual polynomials**. This is the set of polynomials of degree k with $p(0) = 1$. Since the kth GMRES iteration satisfies

$$\|\mathbf{b} - \mathbf{A}\mathbf{x}_k\| \le \|\mathbf{b} - \mathbf{A}\mathbf{z}\|$$

for all $\mathbf{z} \in \mathcal{K}_k$, we must have [107]

$$\|\mathbf{r}_k\| = \min_{p \in \mathcal{P}_k} \|p(\mathbf{A})\mathbf{r}_0\|. \tag{3.2}$$

This simple fact can lead to very useful error estimates.

Here, for example, is a convergence result for diagonalizable matrices. $\mathbf{A}$ is **diagonalizable** if there is a nonsingular matrix $\mathbf{V}$ such that

$$\mathbf{A} = \mathbf{V}\Lambda\mathbf{V}^{-1}.$$

Here Λ is a diagonal matrix with the eigenvalues of $\mathbf{A}$ on the diagonal. If $\mathbf{A}$ is a diagonalizable matrix and p is a polynomial, then

$$p(\mathbf{A}) = \mathbf{V}p(\Lambda)\mathbf{V}^{-1}.$$

$\mathbf{A}$ is **normal** if the **diagonalizing transformation** $\mathbf{V}$ is **unitary**. In this case the columns of $\mathbf{V}$ are the eigenvectors of $\mathbf{A}$ and $\mathbf{V}^{-1} = \mathbf{V}^H$. Here $\mathbf{V}^H$ is the complex conjugate transpose of $\mathbf{V}$. The reader should be aware that $\mathbf{V}$ and Λ can be complex even if $\mathbf{A}$ is real.

Theorem 3.1. *Let* $\mathbf{A} = \mathbf{V}\Lambda\mathbf{V}^{-1}$ *be a nonsingular diagonalizable matrix. Let* $\mathbf{x}_k$ *be the* kth *GMRES iterate. Then, for all* $\bar{p}_k \in \mathcal{P}_k$,

$$\frac{\|\mathbf{r}_k\|}{\|\mathbf{r}_0\|} \le \kappa_2(\mathbf{V}) \max_{z \in \sigma(\mathbf{A})} |\bar{p}_k(z)|, \tag{3.3}$$

where $\sigma(\mathbf{A})$ *is the set of eigenvalues of* $\mathbf{A}$.

Proof. Let $\bar{p}_k \in \mathcal{P}_k$. We can easily estimate $\|\bar{p}_k(\mathbf{A})\|$ by

$$\|\bar{p}_k(\mathbf{A})\| \leq \|\mathbf{F}\|\|\mathbf{V}^{-1}\|\|\bar{p}_k(\Lambda)\| \leq \kappa_2(\mathbf{V}) \max_{z \in \sigma(\mathbf{A})} |\bar{p}_k(z)|,$$

as asserted. $\square$

Suppose, for example, that $\mathbf{A}$ is diagonalizable, $\kappa(\mathbf{V}) = 100$, and all the eigenvalues of A lie in a disk of radius 0.1 centered about 1 in the complex plane. Theorem 3.1 implies (using $\bar{p}_k(z) = (1 - z)^k$) that

$$\frac{\|\mathbf{r}_k\|}{\|\mathbf{r}_0\|} \leq 100(0.1)^k = 0.1^{k-2}.$$

Hence, GMRES will reduce the residual by a factor of, say, 10^5 after seven iterations. Since reduction of the residual is the goal of the linear iteration in an inexact Newton method, this is a very useful bound. See [28] for examples of similar estimates when the eigenvalues are contained in a small number of clusters. One objective of preconditioning (see section 3.3) is to change $\mathbf{A}$ to obtain an advantageous distribution of eigenvalues.

3.2.2 ▪ Low-Storage Krylov Methods

If $\mathbf{A}$ is symmetric and positive definite, the conjugate gradient (CG) method [87] has better convergence and storage properties than the more generally applicable Krylov methods. In exact arithmetic the kth CG iteration minimizes

$$(\mathbf{x} - \mathbf{x}^*)^T \mathbf{A}(\mathbf{x} - \mathbf{x}^*) = \mathbf{e}^T \mathbf{A}\mathbf{e} = \|\mathbf{e}\|_A^2$$

over the kth Krylov subspace. The symmetry and positivity can be exploited so that the storage requirements do not grow with the number of iterations.

A tempting idea is to multiply a general system $\mathbf{A}\mathbf{x} = \mathbf{b}$ by $\mathbf{A}^T$ to obtain the **normal equations** $\mathbf{A}^T\mathbf{A}\mathbf{x} = \mathbf{A}^T\mathbf{b}$ and then apply CG to the new problem, which has a symmetric positive definite coefficient matrix $\mathbf{A}^T\mathbf{A}$. This approach, called CGNR, has the disadvantage that the condition number of $\mathbf{A}^T\mathbf{A}$ is the square of that of $\mathbf{A}$, and hence the convergence of the CG iteration can be far too slow. A similar approach, called CGNE, solves $\mathbf{A}\mathbf{A}^T\mathbf{z} = \mathbf{b}$ with CG and then sets $\mathbf{x} = \mathbf{A}^T\mathbf{z}$. Because the condition number is squared and a transpose-vector multiplication is needed, CGNR and CGNE are used far less frequently than the other low-storage methods.

The need for a transpose-vector multiplication is a major problem unless one wants to store the Jacobian matrix. It is simple (see section 3.4.1) to approximate a Jacobian-vector product with a forward difference, but no matrix-free way to obtain a transpose-vector product is known.

Low-storage alternatives to GMRES that do not need a transpose-vector product are available [66, 107, 197] but do not have the robust theoretical properties of GMRES or CG. Aside from GMRES(m), one such low-storage solver, BiCGSTAB [197] can be used in **nsoli.jl**.

We refer the reader to [79, 107, 197] for detailed descriptions of these methods. If you consider BiCGSTAB and or GMRES(m) as solvers, you should be aware that, while both have the advantage of a fixed storage requirement throughout the linear iteration, there are some problems.

BiCGSTAB can **break down**. This means that the iteration will cause a division by zero. This is not an artifact of the floating point number system but is intrinsic to the methods. While GMRES(m) can also fail to converge, that failure will manifest itself as a stagnation in the iteration, not a division by zero or an overflow.

The number of linear iterations that BiCGSTAB needs for convergence can be roughly the same as for GMRES, but each linear iteration needs two matrix-vector products (i.e., two new evaluations of **F** if you use forward difference Jacobian-vector products).

GMRES(m) should be your first choice and is built into **kl_gmres.jl** and therefore into **nsoli.jl** and **ptcsoli.jl**. If, however, you cannot allocate the storage that GMRES(m) needs to perform well, BiCGSTAB as implemented in **kl_bicgstab.jl** is also accessible from **nsoli.jl** and **ptcsoli.jl** and may solve your problem.

If you can store the Jacobian, or can compute a transpose-vector product in an efficient way, and the Jacobian is well-conditioned, applying the CG iteration to the normal equations can be a good idea. While the cost of a single iteration is two matrix-vector products, convergence, at least in exact arithmetic, is guaranteed [79, 107].

3.3 ▪ Preconditioning

Preconditioning the matrix **A** means multiplying **A** from the right, left, or both sides by a **preconditioner P**. One does this with the expectation that systems with the coefficient matrix **AP** or **AP** are easier to solve than those with **A**. Of course, preconditioning can be done in a matrix-free manner. One needs only a function that performs a preconditioner-vector product.

Left preconditioning multiplies the equation **Ax** = **b** on both sides by **P** to obtain the **preconditioned system PAx = Pb**. One then applies the Krylov method to the preconditioned system. If the condition number of **PA** is really smaller than that of **A**, the residual of the preconditioned system will be a better reflection of the error than that of the original system. One would hope so, since the preconditioned residual will be used to terminate the linear iteration.

Right preconditioning solves the system **APy** = **b** with the Krylov method. Then the solution of the original problem is recovered by setting **x** = **Py**. Right preconditioning has the feature that the residual upon which termination is based is the residual for the original problem.

Two-sided preconditioning replaces **A** with $\mathbf{P}_{left}\mathbf{AP}_{right}$.

A different approach, which is integrated into some initial value problem codes [24, 25], is to pretend that the Jacobian is banded, even if it isn't, and to use Jacobian-vector products and a sparse difference method for banded Jacobians to form a banded approximation to the Jacobian. Then one factors the banded approximation and uses that factorization as the preconditioner.

3.3.1 ▪ Preconditioners

This section is not an exhaustive account of preconditioning and is only intended to point the reader to the literature.

Ideally the preconditioner should be close to the inverse of the Jacobian. In practice, one can get away with far less. If your problem is a discretization of an elliptic differential equation, then a fast solver for the high-order term of the differential operator (with the correct boundary conditions) is an excellent preconditioner [134]. If the high-order term is linear, one might be able to compute the preconditioner-vector product rapidly with, for example, a fast transform method (see section 2.7.5) or a multigrid iteration [23]. Multigrid methods exploit the smoothing properties of the classical stationary iterative methods by mapping the equation through a sequence of grids. When multigrid methods are used as a solver, one can often show that a solution can be obtained at a cost of $O(N)$ operations, where N is the number of unknowns. Multigrid implementation is difficult and a more typical application is to use a single multigrid iteration (for the high-order term) as a preconditioner.

Domain decomposition preconditioners [159, 182, 189] approximate the inverse of the high-order term (or the entire operator) by subdividing the geometric domain of the differential operator, computing the inverses on the subdomains, and combining those inverses. When implemented in an optimal way, the condition number of the preconditioned matrix is independent of the discretization mesh size.

Algebraic preconditioners use the sparsity structure of the Jacobian matrix. This is important, for example, for problems that do not come from discretizations of differential or integral equations or for discretizations of differential equations on unstructured grids, which may be generated by computer programs.

An example of such a preconditioner is **algebraic multigrid** (AMG), which is designed for discretized differential equations on unstructured grids [206]. Algebraic multigrid attempts to recover geometric information from the sparsity pattern of the Jacobian and thereby simulate the intergrid transfers and smoothing used in a conventional geometric multigrid preconditioner. The package **AlgebraicMultigrid.jl** [5] implements several variations of AMG.

Another algebraic approach is **incomplete factorization** [167, 168]. Incomplete factorization preconditioners compute a factorization of a sparse matrix but do not store those elements in the factors that are too small or lie outside a prescribed sparsity pattern. These preconditioners require that the Jacobian be stored as a sparse matrix. The Julia packages **ILUZero.jl** [39] and **IncompleteLU.jl** [186] implement incomplete LU factorizations.

3.4 ▪ Computing an Approximate Newton Step

3.4.1 ▪ Jacobian-Vector Products

For nonlinear equations, the Jacobian-vector product is easy to approximate with a forward difference directional derivative. The forward difference directional derivative at x in the direction w is

$$D_h \mathbf{F}(\mathbf{x} : \mathbf{w}) = \begin{cases} 0, & \mathbf{w} = 0, \\ \|\mathbf{w}\| \dfrac{\mathbf{F}(\mathbf{x} + h\mathbf{w}/\|\mathbf{w}\|) - \mathbf{F}(\mathbf{x})}{h}, & \mathbf{w} \neq 0. \end{cases} \tag{3.4}$$

The scaling is important. We first scale $\mathbf{w}$ to be a unit vector and take a numerical directional derivative in the direction $\mathbf{w}/\|\mathbf{w}\|$. If h is roughly the square root of the error in $\mathbf{F}$, we use a difference increment in the forward difference to make sure that the appropriate low-order bits of $\mathbf{x}$ are perturbed. The same scaling was used in the forward difference Jacobian in (2.1).

3.4.2 ▪ Preconditioning for Nonlinear Equations

Our Newton–Krylov solver **nsoli.jl** allows you to apply a preconditioner on either the left or right by a preconditioner-vector product function. This is simpler and more direct than the approach from [111], where the MATLAB code expects you to incorporate preconditioning into $\mathbf{F}$. We explain that approach later in this section.

We will warn you now, and go into details in section 3.5, that only right preconditioning makes sense for ΨTC .

Left or right preconditioning?

Right preconditioning has a significant advantage in Newton–Krylov methods because the residual is the same residual used in the inexact Newton condition. If one uses left preconditioning, one may terminate on a Newton step that fails to satisfy the inexact Newton condition for the unpreconditioned problem and affect the performance of the line search.

Left preconditioning will terminate the linear iteration when $\|\mathbf{PF}'(\mathbf{x})\mathbf{d} + \mathbf{PF}(\mathbf{x})\|$ is small. If $\mathbf{P}$ is a good approximation to $\mathbf{F}'(\mathbf{x}^*)^{-1}$, then

$$\mathbf{d} \approx -\mathbf{PF}'(\mathbf{x}^*)\mathbf{e} \approx -\mathbf{e}$$

and so one might expect a better Newton step which captures the actual error. However, the meaning of the inexact Newton condition for left preconditioning is not completely clear.

On the other hand, if one uses **right preconditioning**, then the linear iteration will terminate when $\|\mathbf{F}'(\mathbf{x})\mathbf{d} + \mathbf{F}(\mathbf{x})\|$ is small, reflecting (3.1) and responding to the problem statement "make the linear residual small," which corresponds exactly to the inexact Newton condition (1.12) and Theorem 1.4 for convergence of inexact Newton methods.

Preconditioning the nonlinear problem

One can sometime code the combination of preconditioner and nonlinear residual more efficiently than doing them separately. To precondition the equation for the Newton step from the left, one would simply apply **nsoli.jl** to the preconditioned nonlinear problem

$$\mathbf{F}_P(\mathbf{x}) = \mathbf{PF}(\mathbf{x}) = 0.$$

The equation for the Newton step for $\mathbf{F}_P$ is

$$\mathbf{F}'_P(\mathbf{x})\mathbf{d} = \mathbf{PF}'(\mathbf{x})\mathbf{d} = -\mathbf{F}_P(\mathbf{x}) = -\mathbf{PF}(\mathbf{x}),$$

which is the left-preconditioned equation for the Newton direction for $\mathbf{F}$.

Left preconditioning makes more sense if one is preconditioning the nonlinear problem. The reason for this is that the line search is applied to the preconditioned problem, which is not the case if one preconditions the linear solver. The examples in [111] make left preconditioning look a bit too good for exactly this reason. We will revisit this issue in section 3.7.2.

In nonlinear right preconditioning we set $\mathbf{x} = \mathbf{Py}$ and solve

$$\mathbf{F}_P(\mathbf{y}) = \mathbf{F}(\mathbf{Py}) = 0$$

with Newton's method. The equation for the direction $\tilde{\mathbf{d}}$ is

$$\mathbf{F}'_P(\mathbf{y})\tilde{\mathbf{d}} = \mathbf{F}'(\mathbf{Py})\mathbf{P}\tilde{\mathbf{d}} = -\mathbf{F}_P(\mathbf{y}) = -\mathbf{F}(\mathbf{Py}),$$

which is exactly the right-preconditioned equation for the step. Hence right preconditioning for the nonlinear problem produces the same iteration as solving the original problem and using right preconditioning for the linear equation for the Newton direction.

Preconditioning for ΨTC is a different story and we cover that in section 3.5.

3.4.3 ▪ Choosing the Forcing Term

One should take care to choose a forcing term that is not too small at the beginning of the iteration. For example, setting $\eta = 10^{-8}$ for the entire iteration will lead to a low number of nonlinear iterations. However, one will be taking far too much time on linear solves early on. If one wants to fix η, the author's favorite approach, a more sensible value is $\eta = 0.1$, which is the default in **nsoli.jl**. Once again, ΨTC is different and a smaller forcing term, especially at the beginning, may be needed to track the dynamics. Having said all that, we must also discuss a very clever alternative for Newton's method.

The Eisenstat–Walker approach from [60] changes the forcing term η in (3.1) as the nonlinear iteration progresses. The formula is complex and motivated by a lengthy story, which we condense from [107]. The overall goal in [60] is to solve the linear equation for the Newton step to just enough precision to make good progress when far from a solution, but also to obtain quadratic convergence when near a solution. One might base a choice of η on residual norms; one way to do this is

$$\eta_n^{Res} = \gamma\|\mathbf{F}(\mathbf{x}_n)\|^2/\|F(\mathbf{x}_{n-1})\|^2,$$

where $\gamma \in (0, 1]$ is a parameter. If η_n^{Res} is bounded away from 1 for the entire iteration, the choice $\eta_n = \eta_n^{Res}$ will do the job, assuming we make a good choice for η_0. To make sure that η_n stays well away from 1, we can simply limit its maximum size. Of course, if η_n is too small in the early stage of the iteration, then the linear equation for the Newton step can be solved to far more precision than is really needed. To protect against **oversolving**, a method of **safeguarding** was proposed in [60] to avoid volatile decreases in η_n. The idea is that if η_{n-1} is sufficiently large, we do not let η_n decrease by too much; [60] suggests limiting the decrease to a factor of η_{n-1}.

After taking all this into account, one finally arrives at [107]

$$\eta_n = \min(\eta_{max}, \max(\eta_n^{Safe}, 0.5\tau_t/\|\mathbf{F}(\mathbf{x}_n)\|)). \tag{3.5}$$

The term

$$\tau_t = \tau_a + \tau_r\|\mathbf{F}(\mathbf{x}_0)\|$$

is the termination tolerance for the nonlinear iteration and is included in the formula to prevent oversolving on the final iteration. η_{max} is an upper limit on the forcing term and

$$\eta_n^{Safe} = \begin{cases} \eta_{max}, & n = 0, \\[2mm] \min(\eta_{max}, \eta_n^{Res}), & n > 0, \gamma\eta_{n-1}^2 \leq 0.1, \\[2mm] \min(\eta_{max}, \max(\eta_n^{Res}, \gamma\eta_{n-1}^2)), & n > 0, \gamma\eta_{n-1}^2 > 0.1. \end{cases} \qquad (3.6)$$

In [60] the choices $\gamma = 0.9$ and $\eta_{max} = 0.9999$ are used. In **nsoli.jl** we set $\gamma = 0.9$. You can specify η_{max} in the call to **nsoli.jl**.

3.5 ▪ Pseudo-transient Continuation

The considerations from sections 2.7.6 and 2.5.3 are important for all choices of linear solver. There are a few more things to think about when using Krylov solvers to compute the step in ΨTC . The first is the forcing term. Unlike Newton's method, when the Eisenstat–Walker method permits one to use a large forcing term early in the iteration, ΨTC must accurately capture the transient dynamics early, and therefore one should choose a small forcing term to do that.

Preconditioning is also different. Recall that the linear equation for the ΨTC step is

$$(\delta_n^{-1}\mathbf{V} + \mathbf{F}'(\mathbf{x}_n))\mathbf{s} = -\mathbf{F}(\mathbf{x}_n)$$

and the inexact Newton condition is

$$\|(\delta^{-1}\mathbf{V} + \mathbf{F}'(\mathbf{x}))\mathbf{s} + \mathbf{F}(\mathbf{x})\| \leq \eta\|\mathbf{F}(\mathbf{x})\|.$$

The coefficient matrix includes δ, and one might think that the preconditioner should be aware of δ as well. We will explore that question later. More importantly, satisfaction of the inexact Newton condition is very important if one wants to follow the dynamics accurately in the early stages of the iteration. Hence one should precondition the linear equation from the right.

The forcing term is also important for following the dynamics in the early stages. Using a forcing term that is too large can cause a jump to a different stable branch than the one you want.

Left preconditioning and nonlinear preconditioning (see section 3.4.2) are not good ideas for ΨTC . There is no reason for the stability of the preconditioned initial value problem to be the same as that for the original version. For example, if one uses nonlinear right preconditioning and replaces $\mathbf{F}(\mathbf{x})$ with $\mathbf{F}(\mathbf{Py})$, the corresponding time-dependent problem would be

$$\mathbf{P}\frac{d\mathbf{y}}{dt} = -\mathbf{V}^{-1}\mathbf{F}(\mathbf{Py}), \ \mathbf{y}(0) = \mathbf{Px}_0, \qquad (3.7)$$

which could have completely different stability properties.

The linear equation for the ΨTC step depends on δ. One may want the preconditioner to do that as well. `ptcsoli.jl` allows you to do that and we give an example in section 3.8.4.

3.6 ▪ What Can Go Wrong?

Any problem from sections 1.9 or 2.5, of course, can arise if you solve linear systems by iteration. There are a few problems that are unique to Newton iterative methods. The symptoms of these problems are unexpectedly slow convergence or even failure/stagnation of the nonlinear iteration.

3.6.1 ▪ Failure of the Inner Iteration

When the linear iteration does not satisfy the inexact Newton condition (3.1) and the limit on linear iterations has been reached, a sensible response is to warn the user and return the step to the nonlinear iteration. Most codes, including **nsoli.jl**, do this. While it is likely that the nonlinear iteration will continue to make progress, convergence is not certain and one may have to allow the linear solver more iterations, use a different linear solver, or, in extreme cases, find enough storage to use a direct solver.

3.6.2 ▪ Loss of Orthogonality

GMRES exploits orthogonality of the Krylov basis to estimate the residual. In finite-precision arithmetic this orthogonality can be lost and the estimate of the residual in the iteration can be poor. The iteration could terminate prematurely because the estimated residual satisfies (3.1) while the true residual does not. This is a much more subtle problem than failure to converge because the linear solver can report success but return an inaccurate and useless step. The GMRES code **kl_gmres.jl** has several orthogonalization methods as options, but **nsoli.jl** and **ptcsoli.jl** take the fastest and most robust of these options, which is classical Gram–Schmidt orthogonalization repeated twice. This is the approach used in most advanced codes such as Trilinos [86], for example.

3.7 ▪ Newton–Krylov Examples

In this section we consider examples of the choice of the forcing term, left vs. right preconditioning, how preconditioning affects ΨTC , and how low-storage Krylov methods affect the nonlinear iteration. We compare the alternatives with residual histories, as we did in Chapter 2, but compare the residual norm not only as a function of the nonlinear iteration count, but also as a function of the number of Jacobian-vector products. This latter comparison better illustrates the linear algebra cost and therefore the effects of preconditioning.

3.7.1 ▪ Choice of the Forcing Term

In this section we explore how the choice of η affects convergence and cost. We use the default **nsoli.jl** choice of a fixed $\eta = 0.1$ and the Eisenstat–Walker choice with two values of η_{max}. In Figure 3.1, we show the results for the H-equation, a well-conditioned problem with a good initial iterate. In this case the choice of forcing term makes little difference.

In Figure 3.2 we show results for the convection-diffusion problem from section 2.7.5 with no preconditioning or right preconditioning. We can exploit the regular grid by using a **fast Poisson solver P** as a preconditioner. Our preconditioner `fish2d.jl` uses the Julia package **FFTW.jl** [102] to solve the discrete form of $-\nabla^2 w = u$ with homogeneous Dirichlet boundary conditions to return **w = Pu**.

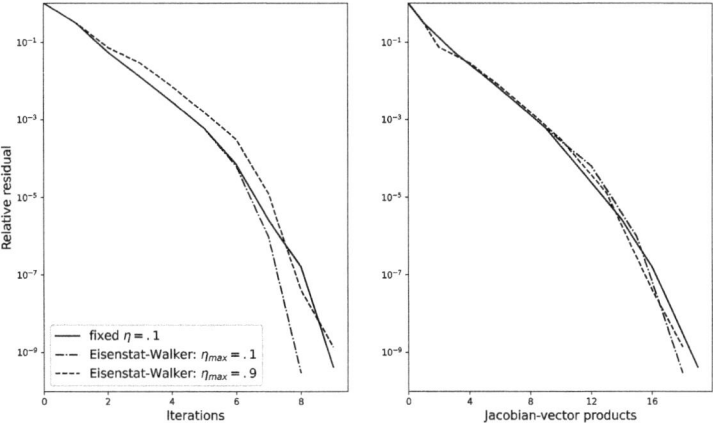

Figure 3.1: Choice of η for the H-equation.

The iteration with the Eisenstat–Walker forcing term with a small value $\eta_{max} = 0.1$ takes fewer nonlinear iterations, but the difference between that choice and fixed η largely vanishes in terms of Jacobian-vector products. This is because Eisenstat–Walker takes more linear iterations in the terminal phase of the nonlinear iteration. Without preconditioning, the larger choice of $\eta = 0.5$ struggles early in the iteration because the large forcing term results in a direction that causes the line search to work harder.

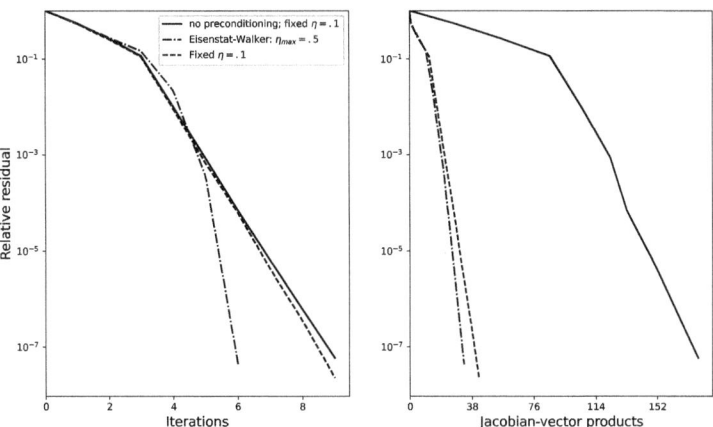

Figure 3.2: Choice of η for the convection-diffusion equation.

3.7.2 ▪ Left vs. Right Preconditioning

We give one example of how right and left preconditioning differ for the convection-diffusion example when we use the Eisenstat–Walker forcing term with $\eta_{max} = 0.9$. This large value of η_{max} can be useful if the initial iterate is poor and an accurate step is not needed, but only if one preconditions from the right. The benchmark results for this example are in the notebook results in section 3.8.3. The results for left preconditioning and $\eta_{max} = 0.9$ are not good.

In this example we precondition the linear equation for the Newton direction with both left and right preconditioning. This means that the line search is based on the unpreconditioned residual. This is a case where left preconditioning can do poorly because it does not directly use the inexact Newton condition for the original nonlinear problem to terminate the linear iteration. As you can see from Figure 3.3 the iteration with left preconditioning spends much more time in the early phase of the iteration. An examination of the `stats.iarm` field in the output shows that the left-preconditioned iteration is in the line search for those iterations. It finally converges, but there is no guarantee of this because the theory is based on satisfying the inexact Newton condition with a value of η that is bounded away from 1, and with left preconditioning the solution of the linearized problem is not guaranteed to do that.

On the other hand, if we precondition the nonlinear problem from the left, then the line search uses the preconditioned residual and the nonlinear convergence is much better. One must be careful in interpreting the figure because the nonlinearly preconditioned residual history (labeled Left-NL) is the preconditioned residual norm and not measuring the same thing as the other two curves. Having said that, the convergence of the nonlinearly preconditioned iteration is fast and takes slightly fewer function evaluations and iterations than the right preconditioned iteration.

The codes for this example are in the file **src/TestProblems/Systems/EllipticPDE.jl**. The nonlinear function for the preconditioned nonlinear problem is **hardleft!**. This function calls the residual for the equation **pdeF!** and follows that with a call to the preconditioner **Pfish2d**, which uses our fast Poisson solver **fish2d.jl**.

```
"""
hardleft!(FV, u, pdata)
Convection-diffusion equation with left preconditioning hard-wired in

"""
function hardleft!(FV, u, pdata)
fdata=pdata.fdata
# Call the nonlinear function
FV = pdeF!(FV,u,pdata)
# and apply the preconditioner.
FV .= Pfish2d(FV,fdata)
return FV
end
```

3.7.3 ▪ Preconditioning Pseudo-transient Continuation

The examples in this section are for the buckling beam. We use the function **FBeam!.jl** for the residual and precondition with a fast (tridiagonal) solver for the high-order term. We consider two choices, both use the tridiagonal matrix $\mathbf{D}_2$ for the second derivative operator homogeneous Dirichlet boundary conditions (see sections 2.7.4 and 2.7.6). We compare preconditioning with an application of a preconditioner that uses the current value of δ, $(\mathbf{D}_2 + \delta^{-1}\mathbf{I})^{-1}$, with one that does not, $\mathbf{D}_2^{-1}$. The functions to apply the preconditioner-vector product are simple. **ptcsoli.jl** allows you to either use δ in the preconditioner or not.

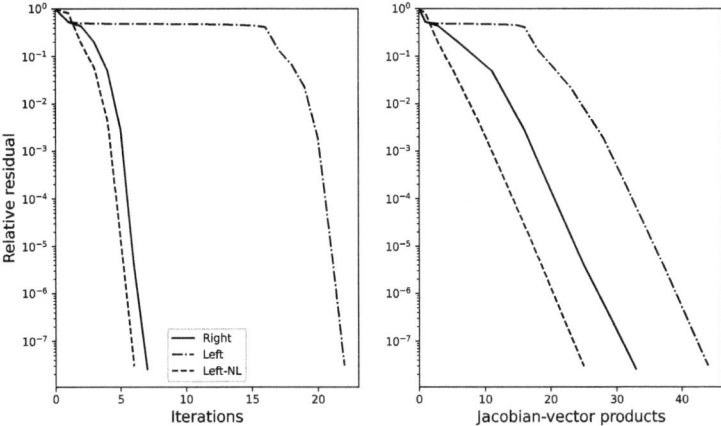

Figure 3.3: Right vs. left for the convection-diffusion equation.

The two functions we list below are contained in **SIAMFANLEquations.jl/src/Examples/ptciBeam.jl** and show one approach to including δ in the preconditioner. The tridiagonal matrix D2 is part of the precomputed data for this problem, and the step `delta` is stored in a preallocated array to make it easier to modify within the solver. We store preallocated data in a named tuple in Julia. This is an immutable structure and one cannot change scalars in such a structure. However, one can put an array in a named tuple and change its contents. This is how we manage to inform the preconditioner of the pseudo-time step `delta`:

```
"""
ptvbeampdt(v, x, bdata)

Precondition buckling beam problem with delta-aware preconditioner.
"""
function ptvbeampdt(v, x, bdata)
    delta = bdata.pdtval[1]
    J = bdata.D2 + (1.0 / delta) * I
    ptv = J \ v
end

"""
ptvbeamp(v, x, bdata)

Precondition buckling beam problem with inverse of high-order term.
"""
function ptvbeam(v, x, bdata)
    J = bdata.D2
    ptv = J \ v
end
```

In Figure 3.4 we plot the residual histories using right preconditioning with both choices of preconditioners and the one from Chapter 2 that uses a direct tridiagonal solve. As you can see, letting the preconditioner know about δ makes little difference in the results.

We advise against left preconditioning for several reasons, and this example illustrates one of them. Here we apply left preconditioning to the buckling beam problem. In Figure 3.5

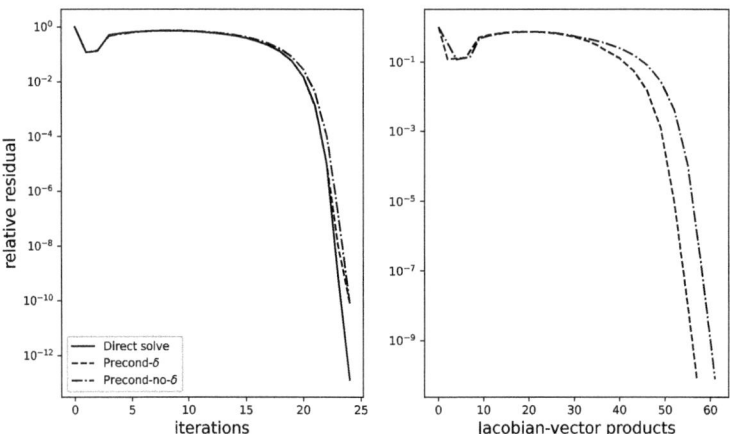

Figure 3.4: Preconditioners for ΨTC.

we plot the residual histories on the left and the solutions on the right. While the residual histories look fine for both preconditioners, the solutions are not the same! The preconditioner that is not aware of the value of δ fails to find the nonnegative solution which is the correct limit of the time-dependent solution. While the nonpositive solution is a stable steady state, it is not the correct solution for the initial data. The preconditioner has let the iteration drift onto the incorrect trajectory. The fact that the preconditioner that is aware of δ found the correct trajectory is likely an accident.

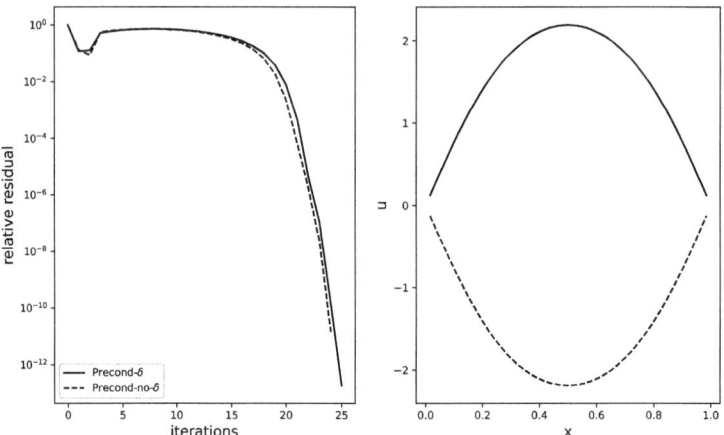

Figure 3.5: Left preconditioning for ΨTC.

3.7.4 ▪ Low-Storage Methods

If you cannot store enough Krylov vectors for GMRES to converge well, then low-storage Krylov methods could help. None of the problems in this book are large enough or poorly conditioned enough for GMRES to have convergence problems. We'll consider three examples in this section, two well-conditioned ones and a poorly conditioned one.

In all the examples we used the Eisenstat–Walker forcing term with $\eta_{max} = 0.1$. The two low-storage methods are BiCGSTAB and GMRES(2). To inform **nsoli.jl** that we are using GMRES(2) we simply allocate only three vectors and set the kwarg `lmaxit = 40` to take 40 GMRES(2) iterations. In the case of the H-equation example, the call looks like

```
# Set up the solve
n = 100
#
# Preallocated storage
#
u0 = zeros(n)
FS = zeros(n)
# Allocate three vectors for GMRES(2)
FPSS = ones(n, 3)
#
# Get organized
#
c = 0.9
atol = 1.e-10
rtol = 1.e-10
hdata = heqinit(u0, c)
koutgs = nsoli( heqf!, u0, FS, FPSS;
    pdata = hdata, rtol = rtol, atol = atol,
    lmaxit = 40, eta = .1, fixedeta = false,
)
```

We begin with the H-equation example on a 100-point grid with $c = 0.9$. Figure 3.6 shows the residual history plotted against nonlinear iterations and also against Jacobian-vector products. Note that the residual histories on the left are very similar. The plot on the right shows that, as you might expect, GMRES uses fewer Jacobian-vector products. BiCGSTAB, which has two Jacobian-vector products for each linear iteration, and

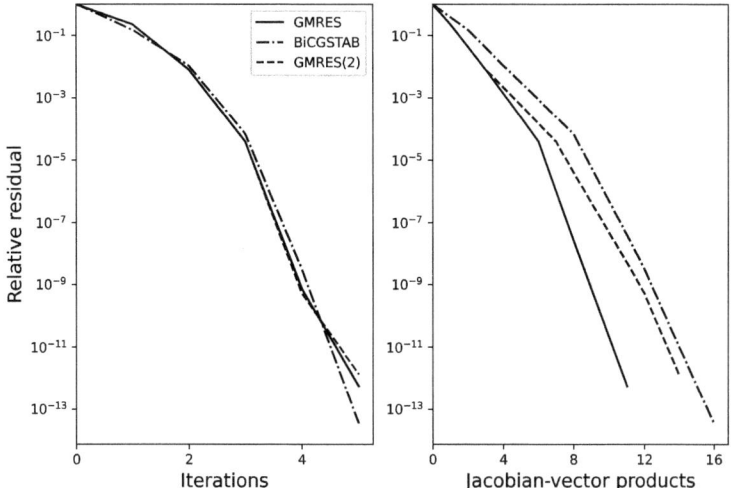

Figure 3.6: Low-storage Krylov methods for the H-equation.

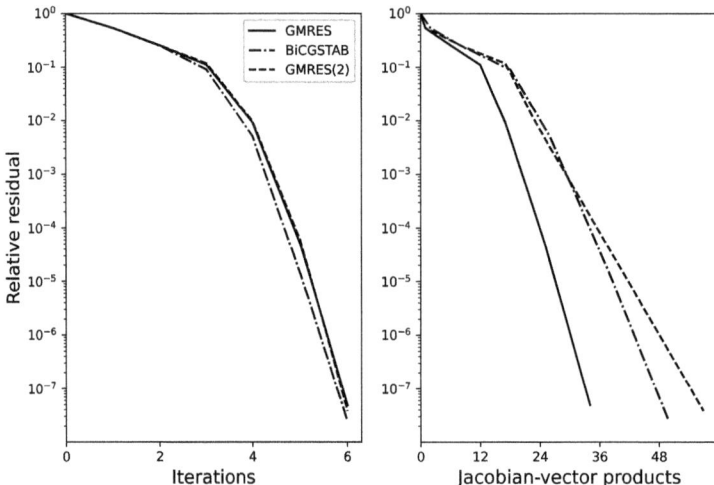

Figure 3.7: Low-storage Krylov methods for the convection-diffusion equation.

GMRES(2), which should take more linear iterations, are not much worse. This problem is very well-conditioned and the initial iterate $u_0 = 1$ is good, so the nonlinear iteration performs well and the linear iteration does not have to work very hard.

Our next example, Figure 3.7, is the convection-diffusion problem. We precondition with the fast Poisson solver as we did in section 3.7.2. We use a 31×31 point mesh and $u_0 = 0$ as the initial iterate. While this problem is well-conditioned, the initial iterate is poor enough to require a line search in the first two iterations.

As was the case for the H-equation, the plot on the left shows that the number of nonlinear iterations was the same for all of the linear solvers. The plot on the right shows that GMRES takes fewer Jacobian-vector products throughout the iteration, both in the line search phase and in the terminal phase where the Eisenstat–Walker forcing term becomes small. Full GMRES is better by roughly 50% than the low-storage methods.

In Figure 3.8 we report on the same experiment using the unpreconditioned convection-diffusion example. The lack of preconditioning significantly increases the cost of the solve, as the plot on the right clearly shows. The performance of GMRES(2) is poor enough to even affect the nonlinear iteration, whereas the nonlinear iteration with full GMRES and BiCGSTAB is similar to the preconditioned case. One can see on the plot on the left that GMRES(2) has trouble finding a good Newton direction in the early phase of the iteration and converges more slowly in the terminal phase. In fact, the GMRES(2) iteration failed to satisfy the inexact Newton condition five times, twice in the early phase and three times in the terminal phase. One can see this by looking at the **stats.ikfail** field in the output. The warning message

```
Linear solver did not meet termination criterion at least once.
     This does not mean the nonlinear solver will fail. lmaxit= 100
```

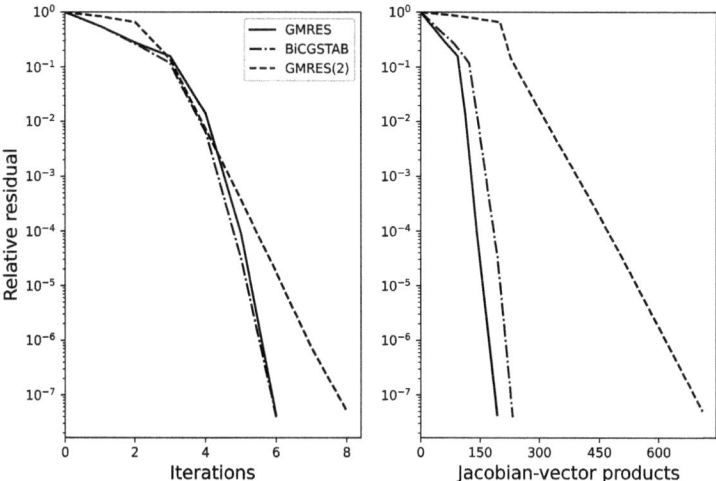

Figure 3.8: Low-storage Krylov methods for the unpreconditioned convection-diffusion equation.

only appears once when you run the solver. As you can see, the nonlinear iteration did converge, but the failure of the inexact Newton condition did cost a few iterations.

3.8 ▪ Notebook: Solvers for Chapter 3

We will follow the pattern of the previous chapters and present two solvers: a Newton code and a ΨTC code. Both codes are for systems of equations and use Krylov methods to compute the step. We have two Krylov solvers: GMRES and BiCGstab.

3.8.1 ▪ nosli.jl

nsoli.jl solves systems of nonlinear equations with Newton–Krylov methods. As usual, we begin with the docstrings.

```
[2]: ?nsoli
```

```
[2]: nsoli(F!, x0, FS, FPS, Jvec=dirder; rtol=1.e-6, atol=1.e-12,
             maxit=20, lmaxit=-1, lsolver="gmres", eta=.1,
             fixedeta=true, Pvec=nothing, pside="right",
             armmax=10, dx = 1.e-7, armfix=false, pdata = nothing,
             printerr = true, keepsolhist = false,
             Krylov_Data = nothing, stagnationok=false)
     )

     C. T. Kelley, 2022

     Julia versions of the nonlinear solvers from my SIAM books.
```

Herewith: nsoli

You must allocate storage for the function and the Krylov basis
in advance -> in the calling program <- ie. in FS and FPS

Inputs:

- F!: function evaluation, the ! indicates that F!
  overwrites FS, your preallocated storage for the function.

  So FS=F!(FS,x) or FS=F!(FS,x,pdata) returns FS=F(x)

  Your function MUST have -> return FS <- at the end. See the
  examples in the docstrings

- x0: initial iterate

- FS: Preallocated storage for function. It is a vector of
  size N

  You should store it as (N) and design F! to use vectors of
  size (N). If you use (N,1) consistently instead, the solvers
  may work, but I make no guarantees.

- FPS: preallocated storage for the Krylov basis. It is
  an N x m matrix where you plan to take at most m-1 GMRES
  iterations before a restart.

- Jvec: Jacobian-vector product. If you leave this out the
  default is a finite difference directional derivative.

  So, FP=Jvec(v,FS,x) or FP=Jvec(v,FS,x,pdata) returns
  FP=F'(x) v.

  (v, FS, x) or (v, FS, x, pdata) must be the argument list,
  even if FP does not need FS. One reason for this is that the
  finite-difference derivative does and that is the default in
  the solver.

- Precision: Lemme tell ya 'bout precision. I designed this
  code for full precision functions and linear algebra in
  any precision you want. You can declare FPS as Float64 or
  Float32 and nsoli will do the right thing. Float16 support
  is there, but not working well.

  If the Jacobian is reasonably well conditioned, you can
  cut the cost of orthogonalization and storage (for GMRES)
  in half with no loss. There is no benefit if your linear
  solver is not GMRES or if orthogonalization and storage of
  the Krylov vectors is only a small part of the cost of the
  computation. So if your preconditioner is good and you only
  need a few Krylovs/Newton, reduced precision won't help you
  much.

BiCGSTAB does not benefit from reduced precision.

Keyword Arguments (kwargs):

rtol and atol: relative and absolute error tolerances

maxit: limit on nonlinear iterations

lmaxit: limit on linear iterations. If lmaxit > m-1, where
FPS has m columns, and you need more than m-1 linear iterations,
then GMRES will restart.

The default is -1 for GMRES. This means that you'll take m-1
iterations, where size(V) = (n,m), and get no restarts. For
BiCGSTAB the default is 10.

lsolver: the linear solver, default = "gmres"

Your choices will be "gmres" or "bicgstab". However, gmres is
the only option for now.

eta and fixed eta: eta > 0 or there's an error

The linear solver terminates when ||F'(x)s + F(x) || <= etag ||
F(x) ||

where

etag = eta if fixedeta=true

etag = Eisenstat-Walker as implemented in book if fixedeta=false

The default, which may change, is eta=.1, fixedeta=true

Pvec: Preconditioner-vector product. The rules are similar to
Jvec So, Pv=Pvec(v,x) or Pv=Pvec(v,x,pdata) returns P(x) v where
P(x) is the preconditioner. You must use x as an input even if
your preconditioner does not depend on x.

pside: apply preconditioner on pside, default = "right". I do
not recommend "left". See Chapter 3 for the story on this.

armmax: upper bound on step size reductions in line search

dx: default = 1.e-7

difference increment in finite-difference derivatives
h=dx*norm(x,Inf)+1.e-8

armfix: default = false

The default is a parabolic line search (ie false). Set to true
and the step size will be fixed at .5. Don't do this unless you
are doing experiments for research.

pdata:

precomputed data for the function, Jacobian-vector, and Preconditioner-vector products. Things will go better if you use this rather than hide the data in global variables within the module for your function/Jacobian.

If you use pdata in any of F!, Jvec, or Pvec, you must use it in all of them.

printerr: default = true

I print a helpful message when the solver fails. To suppress that message set printerr to false.

keepsolhist: default = false

Set this to true to get the history of the iteration in the output tuple. This is on by default for scalar equations and off for systems. Only turn it on if you have use for the data, which can get REALLY LARGE.

Krylov_Data: default = nothing

This is a structure where I put the internal storage for the solvers. You can (but probably should not) preallocate this yourself with the nkl_init function.

Krylov_Data

= nkl_init(n,lsolver)

This is a dangerous thing to mess with and I only recommend it if the allocations in nsoli become a problem in continuation or IVP integration. Krylov_Data is where I store the solution at the end of the iteration and if you reuse it without copying the solution to somewhere else, you'll lose it and it will be overwritten with the new solution. The continuation case study uses this and you should look at that to see what I did.

stagnationok: default = false

Set this to true if you want to disable the line search and either observe divergence or stagnation. This is only useful for research or writing a book.

Output:

  • A named tuple (solution, functionval, history, stats, idid, errcode, solhist)

where

- solution = converged result

- functionval = F(solution)

- history = the vector of residual norms ($||F(x)||$) for the
iteration

- stats = named tuple of the history of (ifun, ijac,
iarm, ikfail), the number of functions/Jacobian-vector
prods/steplength reductions/linear solver failures at each
iteration. Linear solver failures DO NOT mean that the
nonlinear solver will fail. You should look at this stat if,
for example, the line search fails. Increasing the size of FPS
and/or lmaxit might solve the problem.

I do not count the function values for a finite-difference
derivative because they count toward a Jacobian-vector product.

- idid=true if the iteration succeeded and false if not.

- errcode = 0 if the iteration succeeded

  = -1 if the initial iterate satisfies the termination criteria

  = 10 if no convergence after maxit iterations

  = 1 if the line search failed

- solhist:

  This is the entire history of the iteration if you've set
  keepsolhist=true

solhist is an N x K array where N is the length of x and K is
the number of iteration + 1. So, for scalar equations, it's a
row vector.

Example for nsoli

Simple 2D problem. You should get the same results as for
nsol.jl because GMRES will solve the equation for the step
exactly in two iterations. Finite difference Jacobians and
analytic Jacobian-vector products for full precision and finite
difference Jacobian-vector products for single precision.

BiCGSTAB converges in 5 iterations and each nonlinear iteration
costs two Jacobian-vector products. Note that the storage for
the Krylov space in GMRES (jvs) is replace by a single vector
(fpv) when BiCGSTAB is the linear solver.

```julia
julia> function f!(fv,x)
       fv[1]=x[1] + sin(x[2])
       fv[2]=cos(x[1]+x[2])
       return fv
       end
```

```
f! (generic function with 1 method)

julia> function JVec(v, fv, x)
       jvec=zeros(2);
       p=-sin(x[1]+x[2])
       jvec[1]=v[1]+cos(x[2])*v[2]
       jvec[2]=p*(v[1]+v[2])
       return jvec
       end
JVec (generic function with 1 method)

julia> x0=ones(2); fv=zeros(2); jv=zeros(2,2);

julia> jv32=zeros(Float32,2,2);

julia> jvs=zeros(2,3); jvs32=zeros(Float32,2,3);

julia> nout=nsol(f!,x0,fv,jv; sham=1);

julia> kout=nsoli(f!,x0,fv,jvs,JVec;
                  fixedeta=true, eta=.1, lmaxit=2);

julia> kout32=nsoli(f!,x0,fv,jvs32;
                  fixedeta=true, eta=.1, lmaxit=2);

julia> [nout.history kout.history kout32.history]
5×3 Array{Float64,2}:
 1.88791e+00   1.88791e+00   1.88791e+00
 2.43119e-01   2.43120e-01   2.43119e-01
 1.19231e-02   1.19231e-02   1.19230e-02
 1.03266e-05   1.03261e-05   1.03264e-05
 1.46388e-11   1.40862e-11   1.39825e-11

julia> fpv=zeros(2);

julia> koutb=nsoli(f!,x0,fv,fpv,JVec;
              fixedeta=true, eta=.1, lmaxit=2, lsolver="bicgstab");

julia> koutb.history
6-element Vector{Float64}:
 1.88791e+00
 2.43120e-01
 1.19231e-02
 4.87500e-04
 7.54236e-06
 3.84646e-07
```

3.8.2 ▪ Benchmarking the H-Equation with nsoli.jl

We will begin by comparing the fastest solution from Chapter 2 with two variants of Newton-GMRES, one with fixed $\eta = 0.1$ and one with the Eisenstat–Walker forcing term with $\eta_{max} = 0.9$ and $\gamma = 0.9$. I'll allocate 20 vectors for the Krylov basis in the array FPK. First we will compare the residual histories for a small version of the problem.

```
[3]: n=512;
     FS=ones(n,); FPS=ones(n,n); FPS32=ones(Float32,n,n);
         x0=ones(n,); c=.5; hdata = heqinit(x0, c);
     bargs=(atol = 1.e-10, rtol = 1.e-10, sham = 5,
         resdec = .1, pdata=hdata);
     FPK=zeros(n,20);
     # Fixed eta = .1
     kbargs=(atol = 1.e-10, rtol = 1.e-10, eta=.1,
         fixedeta=true, pdata=hdata);
     # Eisenstat-Walker
     kbargsew=(atol = 1.e-10, rtol = 1.e-10, eta=.9,
         fixedeta=false, pdata=hdata);
```

We'll run the winner from Chapter 2 and **nsoli.jl** with two choices for the forcing term.

```
[4]: nout=nsol(heqf!, x0, FS, FPS32, heqJ!; bargs...);
     kout=nsoli(heqf!, x0, FS, FPK; kbargs...);
     koutew=nsoli(heqf!, x0, FS, FPK; kbargsew...);
```

They residual histories are essentially the same. The nonlinear residuals are a bit larger in the middle of the iteration for the Newton–Krylov iterations because of the tolerances for the linear iteration.

```
[5]: [nout.history kout.history koutew.history]
```

```
[5]: 6×3 Matrix{Float64}:
      3.49504e+00   3.49504e+00   3.49504e+00
      1.79698e-02   4.98627e-02   4.98627e-02
      1.55514e-04   1.84641e-03   1.84641e-03
      1.33170e-06   1.82364e-04   1.82364e-04
      1.13964e-08   2.34292e-06   2.34292e-06
      9.75290e-11   2.42545e-11   2.42545e-11
```

Comparing the costs is harder. While a Jacobian-vector product for this problem has the same cost as a call to the function, the cost per iteration for nsol.jl is harder to evaluate in these terms. It's better to look at the benchmark results for a larger problem.

```
[6]: n=4096;
     FS=ones(n); FPS=ones(n,n); FPS32=ones(Float32,n,n);
         x0=ones(n); c=.5; hdata = heqinit(x0, c);
     bargs=(atol = 1.e-10, rtol = 1.e-10, sham = 5,
         resdec = .1, pdata=hdata);
```

```
FPK=zeros(n,20);
kbargs=(atol = 1.e-10, rtol = 1.e-10, eta=.1,
    fixedeta=true, pdata=hdata);
kbargsew=(atol = 1.e-10, rtol = 1.e-10, eta=.9,
    fixedeta=false, pdata=hdata);
```

[7]:
```
println("Shamanskii, n=5");
@btime nsol(heqf!, $x0, $FS, $FPS32, heqJ!; bargs...);
println("Newton-GMRES, fixed eta");
@btime nsoli(heqf!, $x0, $FS, $FPK; kbargs...);
println("Newton-GMRES, Eisenstat-Walker");
@btime nsoli(heqf!, $x0, $FS, $FPK; kbargsew...);
```

```
Shamanskii, n=5
  81.478 ms (8267 allocations: 1.10 MiB)
Newton-GMRES, fixed eta
  1.827 ms (340 allocations: 514.23 KiB)
Newton-GMRES, Eisenstat-Walker
  1.825 ms (340 allocations: 514.23 KiB)
```

The Newton–Krylov code is over 40 times faster. This is not unique to this problem. If your Jacobian is well-conditioned or you have a good preconditioner, as we do in the PDE example, Newton–Krylov should perform much better than any variation of Newton's method using direct linear solvers.

The other interesting thing in this example is that the two forcing term choices performed equally well.

Finally we will see if storing the Krylov basis in single precision improves matters. It's easy to do this by simply replacing the double precision storage for the Krylov basis FPK with the single precision one FPK32.

[8]:
```
n=4096; FS=ones(n);
FPK32=zeros(Float32,n,20)
println("Newton-GMRES, fixed eta");
@btime nsoli(heqf!, $x0, $FS, $FPK32; kbargs...);
println("Newton-GMRES, Eisenstat-Walker");
@btime nsoli(heqf!, $x0, $FS, $FPK32; kbargsew...);
```

```
Newton-GMRES, fixed eta
  1.823 ms (407 allocations: 510.06 KiB)
Newton-GMRES, Eisenstat-Walker
  1.821 ms (407 allocations: 510.06 KiB)
```

There is essentially no difference in performance between storing the basis in single and double precision. It is easy in hindsight to see why. Each function evaluation and forward difference Jacobian-vector product is $O(N \log N)$ work. The cost of orthogonalization for k GMRES iterations with classical Gram–Schmidt twice is $k^2 N$ (can you see why?). So if we do k Krylov iterations per Newton, the cost of orthogonalization is $k^2 N$ and the cost of calls to the residual is $O(kN \log N)$. The computation is dominated by the calls to the residual unless k is very large.

We will quantify this with a computation to look at the iteration statistics. The `ijac` field counts Jacobian-vector projects for each iteration. It is sufficient to look at the fixed $\eta = 0.1$ case. The results for the Eisenstat–Walker forcing term are exactly the same.

```
[9]: fixedetaout = nsoli(heqf!, x0, FS, FPK; kbargs...);
     println(fixedetaout.stats.ijac)
```

```
[0, 1, 1, 1, 1, 2]
```

The statistics indicate that we converge after a single GMRES iteration and are taking a single Krylov per Newton for most of the iteration (remember that the initial iteration is **s** $= 0$ when computing the Newton step). So the orthogonalization cost is N and the function evaluation cost is $O(N \log N)$. We would expect that storing the Krylov basis in single precision would have very little benefit, and that is exactly what we see.

We invite the reader to increase c and the dimension of the problem to see if anything changes.

3.8.3 ▪ Preconditioning the Convection-Diffusion Equation

In this section we will benchmark the Newton-GMRES iteration against the direct solvers from Chapter 2 and explore the differences between left and right preconditioning. We will begin by repeating the computation for the fastest version using **nsol.jl**.

```
[10]: n=31;
      # Get some room for the residual
      u0=zeros(n*n,);
      FV=copy(u0);
      # Get the precomputed data from pdeinit
      pdata=pdeinit(n)
      # Storage for the Jacobian, same sparsity pattern
      # as the discrete Laplacian
      J=copy(pdata.D2);
      # Iteration Parameters
      rtol=1.e-7
      atol=1.e-10
      println("nsol, sham=5");
      @btime nsol(pdeF!, $u0, $FV, $J, pdeJ!;
          resdec=.5, rtol=rtol, atol=atol, pdata=pdata, sham=5);
```

```
nsol, sham=5
  6.483 ms (383 allocations: 6.55 MiB)
```

Now we'll set up the problem for nsoli. We need to allocate storage for the Krylov basis. One case will be no preconditioning at all, so the Krylov basis will need more storage. The analytic Jacobian-vector product is **Jvec2d.jl**, which is in **TestProblems/EllipticPDE.jl**. The preconditioner is **Pvec2d.jl** from **TestProblems/PDE_Tools.jl**.

```
[11]:  # Storage for the Krylov basis
           JV = zeros(n * n, 100)
           eta=.1
           fixedeta=false
       println("nsoli, not preconditioned")
       @btime nsoli(pdeF!, $u0, $FV, $JV, Jvec2d;
                   rtol=rtol, atol=atol, Pvec=nothing, pdata=pdata,
       ↪eta=eta,
                   fixedeta=fixedeta, pside="right");
```

```
nsoli, not preconditioned
  3.426 ms (3942 allocations: 988.25 KiB)
```

Even with no preconditioning, the iterative solver is almost twice as fast as **nsol.jl** using the direct method. When you precondition, which we will do from the right for now, the difference is a factor of almost two over the solve without preconditioning. This difference would increase with a finer mesh. Try it.

```
[12]:  println("nsoli, preconditioned,
           Eisenstat-Walker forcing term")
       @btime nsoli(pdeF!, $u0, $FV, $JV, Jvec2d;
                   rtol=rtol, atol=atol, Pvec=Pvec2d,
                   pdata=pdata, eta=eta,
                   fixedeta=fixedeta, pside="right");
```

```
nsoli, preconditioned,
    Eisenstat-Walker forcing term
  1.954 ms (932 allocations: 659.58 KiB)
```

We will benchmark with a fixed forcing term for our next example.

```
[13]:  fixedeta=true;
       println("nsoli, preconditioned, fixed eta")
       @btime nsoli(pdeF!, $u0, $FV, $JV, Jvec2d;
                   rtol=rtol, atol=atol, Pvec=Pvec2d,
                   pdata=pdata, eta=eta,
                   fixedeta=fixedeta, pside="right");
```

```
nsoli, preconditioned, fixed eta
  2.502 ms (1186 allocations: 935.34 KiB)
```

For this example, we see that Eisenstat–Walker is a bit better. Finally, we return to Eisenstat–Walker with $\eta_{max} = 0.9$. We see very little difference from $\eta_{max} = 0.1$.

```
[14]:  eta=.9; fixedeta=false;
       println("nsoli, preconditioned,
           Eisenstat-Walker forcing term")
       @btime nsoli(pdeF!, $u0, FV, $JV, Jvec2d;
                   rtol=rtol, atol=atol, Pvec=Pvec2d,
```

```
                    pdata=pdata, eta=eta,
                    fixedeta=fixedeta, pside="right");
```

```
nsoli, preconditioned,
    Eisenstat-Walker forcing term
    1.964 ms (958 allocations: 747.75 KiB)
```

Left preconditioning? We'll see that even with $\eta_{max} = 0.1$ it's a bit slower than right preconditioning.

[15]:
```
eta=.1
fixedeta=false
println("nsoli, left preconditioned,
    Eisenstat-Walker forcing term")
@btime nsoli(pdeF!, $u0, $FV, $JV, Jvec2d;
            rtol=rtol, atol=atol, Pvec=Pvec2d,
            pdata=pdata, eta=eta,
            fixedeta=fixedeta, pside="left");
```

```
nsoli, left preconditioned,
    Eisenstat-Walker forcing term
    2.059 ms (1068 allocations: 753.72 KiB)
```

Now we try left preconditioning with $\eta_{max} = 0.9$. We plotted the results in Figure 3.3. While the number of nonlinear iterations is roughly double that of the right-preconditioned version, the solver time is less than the number of nonlinear iterations would indicate. Can you figure out why that is?

Note that we have to increase maxit to give the nonlinear solver enough iterations to overcome the mismatch between left preconditioning and the inexact Newton condition.

[16]:
```
eta=.9;
@btime nsoli(pdeF!, $u0, $FV, $JV, Jvec2d;
            rtol=rtol, atol=atol, Pvec=Pvec2d,
            pdata=pdata, eta=eta, maxit=100,
            fixedeta=fixedeta, pside="left");
```

```
    3.460 ms (1968 allocations: 2.06 MiB)
```

3.8.4 ▪ ptcsoli.jl

ptcsoli.jl is our Newton–Krylov ΨTC code. Herewith the docstrings.

[17]:
```
?ptcsoli
```

[17]:
```
function ptcsoli( F!, x0, FS, FPS, Jvec = dirder; rtol = 1.e-6,
atol = 1.e-12, maxit = 20, lmaxit = -1, lsolver = "gmres", eta
= 0.1, fixedeta = true, Pvec = nothing, PvecKnowsdelta = false,
pside = "right", delta0 = 1.e-6, dx = 1.e-7, pdata = nothing,
```

```
printerr = true, keepsolhist = false, )
```

C. T. Kelley, 2022

Julia versions of the nonlinear solvers from my SIAM books. New for this book ==> ptcsoli

PTC finds the steady-state solution of u' = -F(u), u(0) = u_0. The - sign is a convention.

You must allocate storage for the function and Krylov basis in advance -> in the calling program <- ie. in FS and FPS

Inputs:

- F!: function evaluation, the ! indicates that F! overwrites FS, your preallocated storage for the function.

  So, FS=F!(FS,x) or FS=F!(FS,x,pdata) returns FS=F(x)

  Your function MUST have -> return FS <- at the end. See the example in TestProblems/Systems/FBeam!.jl

- x0: initial iterate

- FS: Preallocated storage for function. It is a vector of size N

  You should store it as (N) and design F! to use vectors of size (N). If you use (N,1) consistently instead, the solvers may work, but I make no guarantees.

- FPS: preallocated storage for the Krylov basis. It is an N x m matrix where you plan to take at most m-1 GMRES iterations before a restart.

- Jvec: Jacobian vector product, If you leave this out the default is a finite difference directional derivative.

  So, FP=Jvec(v,FS,x) or FP=Jvec(v,FS,x,pdata) returns FP=F'(x) v.

  (v, FS, x) or (v, FS, x, pdata) must be the argument list, even if FP does not need FS. One reason for this is that the finite-difference derivative does and that is the default in the solver.

- Precision: Lemme tell ya 'bout precision. I designed this code for full precision functions and linear algebra in any precision you want. You can declare FPS as Float64 or Float32 and ptcsoli will do the right thing. Float16 support is there, but not working well.

  If the Jacobian is reasonably well conditioned, you can cut the cost of orthogonalization and storage (for GMRES)

in half with no loss. There is no benefit if your linear
solver is not GMRES or if orthogonalization and storage of
the Krylov vectors is only a small part of the cost of the
computation. So if your preconditioner is good and you only
need a few Krylovs/Newton, reduced precision won't help you
much.

BiCGSTAB does not benefit from reduced precision.

Keyword Arguments (kwargs):

rtol and atol: relative and absolute error tolerances

delta0: initial pseudo time step. The default value of 1.e-3
is a bit conservative and is one option you really should play
with. Look at the example where I set it to 1.0!

maxit: limit on nonlinear iterations, default=100.

This is coupled to delta0. If your choice of delta0 is too
small (conservative) then you'll need many iterations to
converge and will need a larger value of maxit

For PTC you'll need more iterations than for a straight-up
nonlinear solve. This is part of the price for finding the
stable solution.

lmaxit: limit on linear iterations. If lmaxit > m-1, where
FPS has m columns, and you need more than m-1 linear iterations,
then GMRES will restart.

The default is -1. For GMRES this means that you'll take m-1
iterations, where size(V) = (n,m), and get no restarts. For
BiCGSTAB you'll then get the default of 10 iterations.

lsolver: the linear solver, default = "gmres"

Your choices will be "gmres" or "bicgstab". However, gmres is
the only option for now.

eta and fixed eta: eta > 0 or there's an error.

The linear solver terminates when $||F'(x)s + F(x)|| <=$ etag $||$
$F(x)||$

where

etag = eta if fixedeta=true

etag = Eisenstat-Walker as implemented in book if fixedeta=false

The default, which may change, is eta=.1, fixedeta=true

Pvec: Preconditioner-vector product. The rules are similar to
Jvec So, Pv=Pvec(v,x) or Pv=Pvec(v,x,pdata) returns P(x) v where

P(x) is the preconditioner. You must use x as an input even if your preconditioner does not depend on x.

PvecKnowsdelta: If you want your preconditioner-vector product to depend on the pseudo-timestep delta, put an array deltaval in your precomputed data. Initialize it as deltaval = zeros(1,) and let ptcsoli know about it by setting the kwarg PvecKnowsdelta = true ptcsoli will update the value in deltaval with every change to delta with pdata.deltaval[1]=delta so your preconditioner-vector product can get to it.

pside: apply preconditioner on pside, default = "right". I do not recommend "left". The problem with "left" for ptcsoli is that it can fail to satisfy the inexact Newton condition for the unpreconditioned equation, especially early in the iteration and lead to an incorrect result (unstable solution or wrong branch of steady state). See Chapter 3 for the story on this.

dx: default = 1.e-7

difference increment in finite-difference derivatives h=dx*norm(x)+1.e-8

pdata:

precomputed data for the function, Jacobian-vector, and Preconditioner-vector products. Things will go better if you use this rather than hide the data in global variables within the module for your function/Jacobian

If you use pdata in any of F!, Jvec, or Pvec, you must use it in all of them. precomputed data for the function/Jacobian. Things will go better if you use this rather than hide the data in global variables within the module for your function/Jacobian.

printerr: default = true

I print a helpful message when the solver fails. To suppress that message set printerr to false.

keepsolhist: default = false

Set this to true to get the history of the iteration in the output tuple. This is on by default for scalar equations and off for systems. Only turn it on if you have use for the data, which can get REALLY LARGE.

Output:

A named tuple (solution, functionval, history, stats, idid, errcode, solhist) where

solution = converged result functionval = F(solution) history =
the vector of residual norms ($||F(x)||$) for the iteration stats
= named tuple of the history of (ifun, ijac, ikfail), the number
of functions/Jacobian-vector products/linear solver failures at
each iteration.

I do not count the function values for a finite-difference
derivative because they count toward a Jacobian-vector product.

Linear solver failures need not cause the nonlinear iteration to
fail. You get a warning and that is all.

idid=true if the iteration succeeded and false if not.

errcode = 0 if the iteration succeeded

> = -1 if the initial iterate satisfies the termination criteria
> = 10 if no convergence after maxit iterations

solhist:

This is the entire history of the iteration if you've set
keepsolhist=true

solhist is an N x K array where N is the length of x and K is
the number of iteration + 1. So, for scalar equations, it's a
row vector.

Example for ptcsol

The buckling beam problem. You'll need to use TestProblems for
this to work. The preconditioner is a solver for the high order
term.

```
julia> using SIAMFANLEquations.TestProblems

julia> function PreCondBeam(v, x, bdata)
           J = bdata.D2
           ptv = J\v
       end
PreCondBeam (generic function with 1 method)

julia> n=63; maxit=1000; delta0 = 0.01; lambda = 20.0;

julia> # Set up the precomputed data

julia> bdata = beaminit(n, 0.0, lambda);

julia> x = bdata.x; u0 = x .* (1.0 .- x) .* (2.0 .- x);

julia> u0 .*= exp.(-10.0 * u0); FS = copy(u0); FPJV=zeros(n,20);
```

```
julia> pout = ptcsoli( FBeam!, u0, FS, FPJV;
       delta0 = delta0, pdata = bdata, eta = 1.e-2,
       rtol = 1.e-10, maxit = maxit, Pvec = PreCondBeam);

julia> # It takes a few iterations to get there.
       length(pout.history)
25

julia> [pout.history[1:5] pout.history[21:25]]
5×2 Matrix{Float64}:
 6.31230e+01  1.79574e+00
 7.45927e+00  2.65956e-01
 8.73595e+00  6.58220e-03
 2.91937e+01  8.34114e-06
 3.47970e+01  5.06847e-09

julia> # We get the nonnegative steady state.
julia> maximum(pout.solution)
2.19086e+00

julia> # Now use BiCGSTAB for the linear solver

julia> FPJV=zeros(n);

julia> pout = ptcsoli( FBeam!, u0, FS, FPJV;
       delta0 = delta0, pdata = bdata,
       eta = 1.e-2, rtol = 1.e-10, maxit = maxit,
       Pvec = PreCondBeam, lsolver="bicgstab");

julia> # Same number of iterations as GMRES, but each one costs
       double

julia> # the Jacobian-vector products and much less storage

julia> length(pout.history)
25

julia> [pout.history[1:5] pout.history[21:25]]
5×2 Matrix{Float64}:
 6.31230e+01  1.68032e+00
 7.47081e+00  2.35073e-01
 8.62095e+00  5.18260e-03
 2.96495e+01  3.23803e-06
```

3.8.5 ▪ Benchmarking ΨTC with the Buckling Beam Problem

We will set up the beam problem as we did before. Remember that **bdata.D2** is the discrete Laplacian in one space dimension, which we compute within the initialization function

beaminit. We will start with **ptcsol.jl** to remind you what we did before and solve a larger problem to compare using a direct solver with GMRES.

```
[18]: n=1023; lambda=20; delta=.01; maxit=1000;
      bdata = beaminit(n, 0.0, lambda);
      x = bdata.x; u0 = x .* (1.0 .- x) .* (2.0 .- x);
      u0 .*= exp.(-10.0 * u0);
      FS = copy(u0); FPS=copy(bdata.D2); FPJV = zeros(n, 20);
```

We'll benchmark the solve. Remember that FBeam! and BeamJ! are defined in the **Test-Problems** submodule.

```
[19]: @btime ptcsol(FBeam!, $u0, $FS, $FPS, BeamJ!;
                     rtol=1.e-10, pdata=bdata,
                     delta0=delta, maxit=maxit);
```

```
771.225 mus (655 allocations: 2.76 MiB)
```

To test ptcsoli we will use the δ-dependent preconditioner.

```
[20]: function ptvbeamdelta(v, x, bdata)
          delta = bdata.deltaval[1]
          J = bdata.D2 + (1.0 / delta) * I
          ptv = J \ v
      end
```

```
[20]: ptvbeamdelta (generic function with 1 method)
```

```
[21]: @btime ptcsoli(FBeam!, $u0, $FS, FPJV;
           lsolver="gmres", delta0=delta, pdata=bdata,
           lmaxit=19, eta=1.e-2, Pvec=ptvbeamdelta,
           pside="right", PvecKnowsdelta=true, maxit=maxit);
```

```
1.732 ms (2059 allocations: 6.02 MiB)
```

Using the iterative linear solver costs over two times as much as the direct solver. This is no surprise as the application of the preconditioner requires a tridiagonal solve, which is the same cost as solving the equation for the Newton step with a direct method. The buckling beam problem is simply not hard enough to benefit from an iterative linear solver. The reader should try increasing n to see if anything changes, but should keep in mind that one may need to reduce δ_0 as n increases.

3.9 ▪ Projects

3.9.1 ▪ Low-Storage Solvers

Benchmark the solves for the H-equation and the convection-diffusion equation using BiCGSTAB and GMRES(m) for the linear solvers. How do the runtimes and memory allocations compare to full GMRES? How do the runtimes and allocations depend on the dimension and m for GMRES(m)? Do things change for $c = 1$?

3.9.2 ▪ Mesh Independence

An iteration for a discretization of a differential or integral equation is *mesh-independent* if the iteration statistics are independent of the grid. Nonlinear iterations are usually mesh-independent if the discretization is reasonably well done [2]. That is not the case, however, for the linear solves. One can only get mesh independence for the linear solve if the preconditioner is so good that it essentially converts the problem into an integral equation. For the H-equation and the convection-diffusion (both preconditioned and not), vary the grid size and see how the iteration statistics change. Use both full GMRES and the low-storage solvers. You will want to make figures like the ones earlier in this chapter that plot residual norm against both the number of nonlinear iterations and the number of Jacobian-vector products.

3.9.3 ▪ Playing with the Convection Term

Vary the convection term C in the convection-diffusion equation

$$-\nabla^2 u + Cu(u_x + u_y) = f,$$

where you use the boundary conditions and exact solution u^* from Chapter 2. Vary C from $C = 20$ (the choice in our examples) to $C = 1000$ or larger. Does the choice of forcing term affect the iteration? What happens to the linear and nonlinear iteration statistics?

Chapter 4

Fixed Point Problems and Anderson Acceleration

Files for This Chapter

- From the Package repository:

    - Anderson acceleration solver: **/src/Solvers/aasol.jl**

    - Test problems: **/src/TestProblems/Systems**

        * H-equation: **Hequation.jl**

        * Convection-diffusion equation: **EllipticPDE.jl** and **PDE_Tools.jl**

- From the Notebook repository: **/src/Chapter4**
  Julia codes that generate the figures and tables

4.1 ▪ Fixed Point Problems

In this chapter we focus on fixed point problems

$$\mathbf{x} = \mathbf{G}(\mathbf{x}). \tag{4.1}$$

While one might think that there is little difference between fixed point problems and nonlinear equations $\mathbf{F}(\mathbf{x}) = 0$, that would be wrong [143]. While simply replacing $\mathbf{F}(\mathbf{x})$ with $\mathbf{G}(\mathbf{x}) - \mathbf{x}$ and applying a form of Newton's method is reasonable in most cases, that approach can miss important structural properties of $\mathbf{G}$ and requires computing a Jacobian or Jacobian-vector product, which is not always possible. Conversely, applying a fixed point method to a nonlinear equation with the transformation $\mathbf{G}(\mathbf{x}) = \mathbf{x} + \mathbf{F}(\mathbf{x})$ can often lead to failure, even in the linear case. Having said that, there are problems, such as the H-equation from section 2.7.2, that can be expressed equally well as nonlinear equations or fixed point problems.

The fundamental method for solving fixed point problems is Picard or fixed point iteration

$$\mathbf{x}_{n+1} = \mathbf{G}(\mathbf{x}_n). \tag{4.2}$$

Note that, unlike Newton's method, no derivatives are needed in the iteration.

Throughout this chapter we will assume that $\mathbf{G}$ is a contraction on a set $D \subset R^N$. This means that there is $\sigma \in (0, 1)$ such that

$$\|\mathbf{G}(\mathbf{x}) - \mathbf{G}(\mathbf{y})\| \leq \sigma \|\mathbf{x} - \mathbf{y}\| \tag{4.3}$$

for all $\mathbf{x}, \mathbf{y} \in D$. We will call σ the **contractivity constant**.

The following is the **Contraction Mapping Theorem**.

Theorem 4.1. *If $\mathbf{G}$ is a contraction on $D \in R^N$, then there is a unique solution $\mathbf{x}^* \in D$ of (4.1), and for any $\mathbf{x}_0 \in D$ the iteration (4.2) converges to $\mathbf{x}^*$. Moreover the convergence is q-linear,*

$$\|\mathbf{e}_{n+1}\| \leq \sigma \|\mathbf{e}_n\|. \tag{4.4}$$

As in the previous chapters, $\mathbf{e} = \mathbf{x} - \mathbf{x}^*$.

We remind the reader that the uniqueness in Theorem 4.1 means that $\mathbf{x}^*$ is the only solution **in** D. If $\mathbf{G}$ is nonlinear, there could well be other fixed points outside of D.

Note that the theorem does not specify the norm $\| \cdot \|$. In fact if (4.3) holds for any norm, the iteration will converge. One does not have to know what that norm is, as the linear case illustrates. If you measure convergence in a norm other than the one in which $\mathbf{G}$ is a contraction, you will still observe convergence, but not necessarily q-linear convergence. At worst you will see **r-linear convergence**,

$$\|\mathbf{e}_n\| = O(\sigma^n), \tag{4.5}$$

or, equivalently,

$$\limsup_{n \to \infty} \left(\frac{\|\mathbf{e}_n\|}{\|\mathbf{e}_0\|} \right)^{1/n} \leq \sigma. \tag{4.6}$$

It's worthwhile to consider the case of linear equations

$$\mathbf{A}\mathbf{x} = \mathbf{b} \tag{4.7}$$

and linear fixed point problems

$$\mathbf{x} = \mathbf{M}\mathbf{x} + \mathbf{b}. \tag{4.8}$$

In the case of linear fixed point problems, contractivity of $\mathbf{G}(\mathbf{x}) = \mathbf{M}\mathbf{x} + \mathbf{b}$ means that $\|\mathbf{M}\| < 1$ in some operator norm, equivalently $\rho(\mathbf{M}) < 1$. Here ρ is the **spectral radius**

$$\rho(\mathbf{M}) = \max_{\lambda \in \sigma(\mathbf{M})} |\lambda|,$$

and $\sigma(\mathbf{M})$ is the set of eigenvalues of $\mathbf{M}$. The convergence result is the **Banach Lemma**.

Lemma 4.2. *Suppose* $\|\mathbf{M}\| < 1$. *Then for any* $\mathbf{x}_0 \in R^N$ *the iteration*

$$\mathbf{x}_{n+1} = \mathbf{M}\mathbf{x}_n + \mathbf{b}$$

converges to $\mathbf{x}^* = (\mathbf{I} - \mathbf{M})^{-1}\mathbf{b}$. *Moreover*

$$\|\mathbf{e}_{n+1}\| \leq \|\mathbf{M}\|\|\mathbf{e}_n\|.$$

The Banach Lemma differs from the Contraction Mapping Theorem in that the map $\mathbf{G}(\mathbf{x}) = \mathbf{M}\mathbf{x} + \mathbf{b}$ is a contraction everywhere (so $D = R^N$) if $\|\mathbf{M}\| < 1$. The Contraction Mapping Theorem for nonlinear problems is local in the sense that the domain D is part of the assumptions. Similarly to Newton's method, we can connect contractivity to the Jacobian.

Corollary 4.3. *Suppose* $\mathbf{G}$ *is continuously differentiable near a fixed point* $\mathbf{x}^* = \mathbf{G}(\mathbf{x}^*)$ *and that*

$$\|\mathbf{G}'(\mathbf{x}^*)\| < 1.$$

Then there is $\delta > 0$ *such that* $\mathbf{G}$ *is a contraction in the set*

$$D = \{\mathbf{x} \mid \|\mathbf{x} - \mathbf{x}^*\| < \delta\},$$

with contractivity constant

$$\sigma = \|\mathbf{G}'(\mathbf{x}^*)\| + O(\delta).$$

4.1.1 ▪ Damping

Sometimes one can replace a noncontractive $\mathbf{G}$ with

$$\mathbf{G}_\beta(\mathbf{x}) = (1 - \beta)\mathbf{x} + \beta\mathbf{G}(\mathbf{x}) \tag{4.9}$$

so that $\mathbf{G}_\beta$ is a contraction. Typically $-1 \leq \beta \leq 1$. If one applies Picard iteration to $\mathbf{G}_\beta$, the resulting method is called **damped Picard iteration**,

$$\mathbf{x}_{n+1} = (1 - \beta)\mathbf{x}_n + \beta\mathbf{G}(\mathbf{x}_n). \tag{4.10}$$

So damped Picard iteration with $\beta = 1$ is simply Picard iteration.

To see the limitations and possibilities of this approach, it helps to consider the linear case. Suppose, for some $0 < \mu < 1 < K$,

$$\sigma(\mathbf{M}) \subset [-K, \mu]$$

and so $\rho(\mathbf{M}) \geq K > 1$. If we change $\mathbf{M}$ to

$$\mathbf{M}_\beta = (1 - \beta)\mathbf{I} + \beta\mathbf{M},$$

then, for any $0 < \beta < 2/(K + 1)$,

$$\sigma(\mathbf{M}_\beta) \subset [1 - \beta(1 + K), 1 - \beta(1 - \mu)] \subset (-1, 1)$$

and therefore $\rho(\mathbf{M}_\beta) < 1$. Hence the iteration

$$\mathbf{x}_{n+1} = \mathbf{M}_\beta \mathbf{x}_n + \beta \mathbf{b} = (1 - \beta)\mathbf{x}_n + \beta(\mathbf{M}\mathbf{x}_n + \mathbf{b})$$

converges to the solution of

$$\mathbf{x}^* = \mathbf{M}_\beta \mathbf{x}^* + \beta \mathbf{b} = (1 - \beta)\mathbf{x}^* + \beta(\mathbf{M}\mathbf{x}^* + \mathbf{b})$$

$$= \mathbf{x}^* - \beta(\mathbf{x}^* - \mathbf{M}\mathbf{x}^* - \mathbf{b}),$$

and $\mathbf{x}^* = \mathbf{M}\mathbf{x}^* + \mathbf{b}$.

However, the convergence rate is slow if K is large; for example, if $\beta = 1/(K+1)$, then

$$\rho(\mathbf{M}_\beta) \leq 1 - \beta(1 - \mu) = \frac{K + \mu}{K + 1},$$

which can be very close to one if K is large or μ is close to 1. So, while damping can manufacture contractivity in some (**but not all**) cases, the convergence is likely to be poor. Similarly, if

$$\sigma(\mathbf{M}) \subset [-\mu, K],$$

then any $-2/(K+1) < \beta < 0$ will work as a damping parameter. Damping will not always work. If, for example, the eigenvalues of $\mathbf{M}$ are ± 2, no damping parameter will make $\rho(\mathbf{M}_\beta) < 1$.

We now return to the nonlinear case and state an analogue of Corollary 4.3.

Corollary 4.4. *Suppose* $\mathbf{G}$ *is continuously differentiable near a fixed point* $\mathbf{x}^* = \mathbf{B}(\mathbf{x}^*)$ *and that there is* β *such that*

$$\|(1 - \beta)\mathbf{I} + \beta \mathbf{G}'(\mathbf{x}^*)\| = \hat{\sigma} < 1.$$

Then for any $\sigma \in (\hat{\sigma}, 1)$ *there is* $\delta > 0$ *such that* $\mathbf{G}_\beta$ *is a contraction in the set*

$$D = \{\mathbf{x} \mid \|\mathbf{x} - \mathbf{x}^*\| < \delta\},$$

with contractivity constant σ.

4.1.2 ▪ Mapping Nonlinear Equations to Fixed Point Problems

If, and this is usually a lot to ask, you have a very good preconditioner $\mathbf{P}$ for $\mathbf{F}(\mathbf{x}) = 0$, you can apply the preconditioner from the left and express the nonlinear equation as a fixed point problem for

$$\mathbf{G}(\mathbf{x}) = \mathbf{x} - \mathbf{P}\mathbf{F}(\mathbf{x}). \tag{4.11}$$

Right preconditioning is similar and we will focus only on left preconditioning in this section.

In order for a convergence result like Corollary 4.3 to hold we need the standard assumptions for Newton's method and also require that $\mathbf{P}$ be an **approximate inverse** for $\mathbf{F}'(\mathbf{x}^*)$. This means that

$$\|\mathbf{I} - \mathbf{P}\mathbf{F}'(\mathbf{x}^*)\| < 1. \tag{4.12}$$

If (4.12) holds and $\mathbf{G}$ is given by (4.11), then

$$\|\mathbf{G}'(\mathbf{x}^*)\| = \|\mathbf{I} - \mathbf{PF}'(\mathbf{x}^*)\| < 1$$

and we can apply Corollary 4.3 directly. It's worthwhile to state the convergence result explicitly.

Corollary 4.5. *Let the standard assumptions hold and let $\mathbf{P}$ be an approximate inverse of $\mathbf{F}'(\mathbf{x}^*)$. Let $\mathbf{G}$ be given by (4.11). Assume that*

$$\|\mathbf{I} - \mathbf{PF}'(\mathbf{x}^*)\| < \sigma < 1.$$

Then there is $\delta > 0$ such that $\mathbf{G}$ is a contraction in the set

$$D = \{\mathbf{x} \mid \|\mathbf{x} - \mathbf{x}^*\| < \delta\},$$

with contractivity constant σ.

One can view any Newton-like method as a fixed point iteration in light of (4.11). For example, if $\mathbf{P} = \mathbf{F}'(\mathbf{x}_0)^{-1}$, then the fixed point iteration is the chord method. If you allow $\mathbf{P}$ to depend on $\mathbf{x}$, then you can analyze Newton's method itself as a fixed point iteration [149].

4.1.3 ▪ Preconditioning Fixed Point Maps

If we start from a nonlinear equation $\mathbf{F}(\mathbf{x}) = 0$ and an approximate inverse for $\mathbf{F}'(\mathbf{x}^*)$, the resulting fixed point map is given by (4.11). If, however, we start from a fixed point problem $\mathbf{G}(\mathbf{x}) = \mathbf{x}$, then the corresponding nonlinear equation, following the convention from [6, 201], is $\mathbf{F}(\mathbf{x}) = \mathbf{G}(\mathbf{x}) - \mathbf{x}$. This leads to an approach to preconditioning $\mathbf{G}$ by finding a good preconditioner for $\mathbf{F}$ [84]. For example, if one can find an approximate inverse $\mathbf{P}$ for

$$\mathbf{F}'(\mathbf{x}^*) = \mathbf{G}'(\mathbf{x}^*) - \mathbf{I},$$

then one can apply formula (4.11) and transform $\mathbf{G}$ to

$$\mathbf{G}_\mathbf{P}(\mathbf{x}) = \mathbf{x} - \mathbf{PF}(\mathbf{x}) = \mathbf{x} - \mathbf{P}(\mathbf{G}(\mathbf{x}) - \mathbf{x}),$$

which is a special case of equation (18) in [84].

4.2 ▪ Anderson Acceleration

Anderson acceleration [6] is an iterative method for fixed point problems and is intended to accelerate Picard iteration. Anderson acceleration was originally designed for integral equations and is now very common in electronic structure computations (see [50] and many references since then). Anderson acceleration is essentially the same as DIIS (Direct Inversion on the Iterative Subspace) [124, 129, 166, 174], nonlinear GMRES [31, 136, 146, 202], and interface quasi-Newton [51, 81, 130]. It is also closely related to Pulay mixing [157], also known as CDIIS (Commutator DIIS), [71, 93, 95, 166]. Other variations on the method include CROP (Conjugate Residual with OPtimal trial vectors) [61, 210] and EDIIS (Energy DIIS) [33, 124].

Anderson acceleration is an important component of many electronic structure codes. These, in turn, are critical parts of quantum physics and chemistry codes. Three examples are the RMG (Real Space Multigrid) large scale electronic structure computations code [22, 138], the **DFTK.jl** Julia package [84, 85], and the Gaussian [68, 124] code for computational chemistry. Other applications include stiff dislocation dynamics [70], hydrology [131], neutron transport [1, 8, 191, 205], thermal radiation transport [4], fluid mechanics [153, 154], multiphysics coupling [82, 190, 193], and fluid-structure interaction [51, 69, 81, 130].

Anderson acceleration is most useful when Newton's method is not practical because obtaining approximate Jacobians or Jacobian-vector products is too costly or not possible at all. Comparisons indicate that Newton's method usually performs better when even moderately accurate derivative information can be had at reasonable cost [82]. Having said that, there are cases when Anderson acceleration does outperform an efficient implementation of Newton's method [192]. We provide examples of this poorly understood phenomenon in sections 4.4.1 and 4.4.2.

Algorithm **anderson** is the simplest form of the iteration, which is all we cover in this chapter.

ALGORITHM 4.1.
anderson$(\mathbf{x}_0, \mathbf{G}, m, \beta)$
  $\mathbf{x}_1 = \mathbf{G}(\mathbf{x}_0); \mathbf{F}_0 = \mathbf{G}(\mathbf{x}_0) - \mathbf{x}_0$
  for $k = 1, \ldots$ **do**
    Choose $m_k \le \min(m, k)$
    $\mathbf{F}(\mathbf{x}_k) = \mathbf{G}(\mathbf{x}_k) - \mathbf{x}_k$
    Minimize $\| \sum_{j=0}^{m_k} \alpha_j^k \mathbf{F}(\mathbf{x}_{k-m_k+j}) \|$ subject to
    $\sum_{j=0}^{m_k} \alpha_j^k = 1.$
    $\mathbf{x}_{k+1} = (1 - \beta) \sum_{j=0}^{m_k} \alpha_j^k \mathbf{x}_{k-m_k+j} + \beta \sum_{j=0}^{m_k} \alpha_j^k \mathbf{G}(\mathbf{x}_{k-m_k+j})$
  end for

We discuss the convergence results, the application of the algorithm, and the implementation in terms of these algorithmic components and parameters.

- m is the **depth**. Typically in applications m is small [29], say ≤ 10. There is usually little advantage in large m.

- β is the **mixing or damping parameter**. It plays exactly the same role as it does for Picard iteration. One could vary β as the iteration progresses [30] but we will not do that in this chapter.

- $\mathbf{F}(\mathbf{x}) = \mathbf{G}(\mathbf{x}) - \mathbf{x}$ is the **residual**.

- The **optimization problem** is

$$\min \left\| \sum_{j=0}^{m_k} \alpha_j^k \mathbf{F}(\mathbf{x}_{k-m_k+j}) \right\| \text{ subject to } \sum_{j=0}^{m_k} \alpha_j^k = 1.$$

- The **coefficients** are the α_j^k's.

The reader who is familiar with GMRES may notice the similarity between Anderson acceleration and GMRES for linear problems. In fact [201] the iterations for Anderson acceleration and GMRES can be transformed into one another in most cases for linear problems. This does not mean that Anderson acceleration is a good choice for a linear solver. The storage cost, as we will see, is two to three times more than GMRES.

4.2.1 ▪ Algorithmic Details

As you can see from the algorithmic description, Anderson acceleration keeps a history of the iteration. This history has $m + 1$ vectors for prior residuals and $m + 1$ vectors for prior fixed point map evaluations. Hence Anderson(m) stores at least twice as much as GMRES(m) when applied to a linear problem. There are more efficient ways to implement the method than a direct translation of Algorithm **anderson** into code, but the storage requirements do not change. Note that Anderson(0) is Picard iteration.

Similarly to Picard iteration, Anderson acceleration with mixing parameter β is the same iteration as Anderson acceleration with mixing parameter 1 applied to the map $\mathbf{G}_\beta$ defined in (4.9). For this reason we will state our convergence results for $\beta = 1$.

Note that we have not specified the norm in the optimization problem, and the convergence theorem we will present does not depend on the choice of norm. However, the norm has a significant effect on the implementation. If one uses the ℓ^1 or ℓ^∞ norm, then the optimization problem can be formulated as a linear programming problem and solved with many codes [77]. However, it is most efficient to use the ℓ^2 norm and formulate the optimization problem as a linear least squares problem. To do this first solve the linear least squares problem

$$\min \left\| \mathbf{F}(\mathbf{x}_k) - \sum_{j=0}^{m_k-1} \alpha_j^k (\mathbf{F}(\mathbf{x}_{k-m_k+j}) - \mathbf{F}(\mathbf{x}_k)) \right\|_2^2 \qquad (4.13)$$

for $\{\alpha_j^k\}_{j=0}^{m_k-1}$. Then compute $\alpha_{m_k}^k$ by

$$\alpha_{m_k}^k = 1 - \sum_{j=0}^{m_k-1} \alpha_j^k.$$

You should be aware that the least squares problem (4.13) can be very ill-conditioned, as we will see in the examples. The most stable way to solve (4.13) is to maintain a QR factorization of the coefficient matrix for the optimization problem [38, 190, 194, 201]. We will discuss this a bit more in section 4.3.1.

There are production electronic structure codes, for example, the RMG code [22, 138], which solve the optimization problem with the normal equations. While one would think that such an approach would lead to poor results, it does not. The reason for this, as the analysis [192] shows, is that the residual norm for the least squares problem is the important thing, not a highly accurate solution for the coefficients.

Finally, the choice of m_k in the original form of the algorithm was $\min(m, k)$. One could think of adapting m_k to do things like control the conditioning of the least squares problem (4.13) [4, 190, 201], but most implementations do not do that. We have found that overthinking m_k can make convergence slower even when the linear least squares problem is poorly conditioned.

4.3 ▪ Convergence Theory

Convergence analysis has been reported in the literature only recently and most of that work assumes at least continuous differentiability of the fixed point map. The first convergence theorem for the method as used in practice was in [192].

Many papers [63, 155, 166, 169, 201] make connections between Anderson acceleration and multisecant quasi-Newton methods and, for linear problems, GMRES. These connections do not lead to convergence results except in the linear case where $m = N + 1$, that is, to allow GMRES to converge to the solution in N iterations.

There are convergence results for the linear case [46, 192, 201], the continuously differentiable case [33], the Lipschitz-continuously differentiable case [191, 192], and even smoother cases [62, 154]. The results in [62, 154] connect the improvement in the residual from the optimization problem over the residual one would have in Picard iteration ($m = 0$) with the performance of the iteration and quantify the improvement of Anderson acceleration over Picard iteration in certain cases. One can formulate Anderson acceleration as a larger fixed point problem for the most recent $m + 1$ iterations [47] and reveal some fine structure of the iteration by doing that. Convergence results and algorithms for nonsmooth problems can be found in [18, 33, 124, 209]. The assumption on contractivity of $\mathbf{G}$ can also be weakened in certain cases [153].

We will give the version of the convergence theorem from [33] in this section. That result differs from the one in [192] only in that the requirements on differentiability of $\mathbf{G}$ are weaker. Note that we do not assume that the coefficients $\{\alpha_j^k\}$ come from any particular optimization problem, only that the linear combination of residuals has norm no larger than that of the most recent residual. This generality admits variations such as EDIIS [33, 124], for example.

Assumption 4.3.1. $\mathbf{G}$ *is a continuously differentiable contraction on* $D \subset R^N$ *with contractivity constant* σ *and* $\mathbf{x}^*$ *is the unique fixed point of* $\mathbf{G}$ *in* D. *Assume that for each iteration k there are real* $\{\alpha_j^k\}_{j=0}^{m_k}$ *with* $0 \leq m_k \leq \min(m, k)$ *such that*

$$\sum_{j=0}^{m_k} \alpha_j^k = 1,$$

there is M_α *such that*

$$\sum_{j=0}^{m_k} |\alpha_j^k| \leq M_\alpha \tag{4.14}$$

for all $k \geq 0$,

$$\mathbf{x}_{k+1} = \sum_{j=0}^{m_k} \alpha_j^k \mathbf{G}(\mathbf{x}_{k-m_k+j}), \tag{4.15}$$

and

$$\left\| \sum_{j=0}^{m_k} \alpha_j^k \mathbf{F}(\mathbf{x}_{k-m_k+j}) \right\| \leq \|\mathbf{F}(\mathbf{x}_k)\|. \tag{4.16}$$

The convergence result is as follows.

Theorem 4.6. *Let Assumption 4.3.1 hold. Assume that*

$$\|\mathbf{x}_0 - \mathbf{x}^*\| \le \delta.$$

Then if δ is sufficiently small, the iteration given by (4.15) and (4.16) converges to $\mathbf{u}^*$ and

$$\limsup_{k \to \infty} \left(\frac{\|\mathbf{F}(\mathbf{u}_k)\|}{\|\mathbf{F}(\mathbf{u}_0)\|} \right)^{1/k} \le \sigma. \tag{4.17}$$

The assumption that the initial iterate is near the solution is more than a theoretical convenience. In fact Anderson acceleration can fail to converge with a poor initial iterate. One example is electronic structure computations for metallic systems where the HOMO-LUMO gap is small [124]. In such cases one must use a small mixing parameter to ensure convergence. However, a small mixing parameter degrades the performance of the iteration. In addition, an accurate initial iterate is often necessary for convergence, and it is often difficult to find a sufficiently good initial iterate [68,124,172,207,208]. The EDIIS (Energy DIIS) algorithm from [124] was designed to address these problems by adding a nonnegativity constraint to the optimization problem. While there is theory for this method [33], the performance can be poor if Anderson acceleration converges well without modification [33, 124].

4.3.1 ▪ Implementation

While the formulation of the optimization problem in (4.13) is useful for analysis, it is not efficient because the coefficient matrix must be reformed with each iteration. Moreover, if one plans to use a QR factorization to solve the least squares problem, that factorization would also have to be done from scratch with each iteration.

A better approach is to reformulate the optimization problem as

$$\min_{\theta \in R^{m_k}} \left\| \mathbf{F}(\mathbf{x}_k) - \sum_{j=0}^{m_k-1} \theta_j (\mathbf{F}(\mathbf{x}_{k-m_k+j+1}) - \mathbf{F}(\mathbf{x}_{k-m_k+j})) \right\| \tag{4.18}$$

to obtain a vector $\theta^k \in R^{m_k}$. Then the next iteration is

$$\mathbf{x}_{k+1} = \mathbf{G}(\mathbf{x}_k) - \sum_{j=0}^{m_k-1} \theta_j^k (\mathbf{G}(\mathbf{x}_{k-m_k+j+1}) - \mathbf{G}(\mathbf{x}_{k-m_k+j})). \tag{4.19}$$

In terms of (4.13),

$$\alpha_0 = \theta_0, \ \alpha_j = \theta_j - \theta_{j-1} \quad \text{for } 1 \le j \le m_k - 1 \text{ and } \alpha_{m_k} = 1 - \theta_{m_k-1}.$$

Hence, with each iteration one must add (and delete if $k > m$) a column from the matrices $\mathbf{D}_F^k$ and $\mathbf{D}_G^k$ where the columns of $\mathbf{D}_F^k$ and $\mathbf{D}_G^k$ are, for $0 \le j \le m_k - 1$,

$$(\mathbf{D}_F^k)_j = \mathbf{F}(\mathbf{x}_{k-m_k+j+1}) - \mathbf{F}(\mathbf{x}_{k-m_k+j})$$

and

$$(\mathbf{D}_G^k)_j = \mathbf{G}(\mathbf{x}_{k-m_k+j+1}) - \mathbf{G}(\mathbf{x}_{k-m_k+j}).$$

So if $\mathbf{D}_F^k = \mathbf{Q}^k \mathbf{R}^k$ is the QR factorization of $\mathbf{D}_F^k$, then

$$\theta = (\mathbf{R}^k)^{-1}(\mathbf{Q}^k)^T \mathbf{F}(\mathbf{x}_k)$$

and

$$\mathbf{x}_{k+1} = \mathbf{G}(\mathbf{x}_k) - \mathbf{D}_G^k \theta.$$

It should be clear that managing $\mathbf{D}_G^k$ is simply a matter of adding a new column if $k < m - 1$. If $k \geq m - 1$, then one must first delete the leading column ($j = 0$) by shifting columns $(1, \ldots, m - 1)$ to columns $(0, \ldots, m - 1)$ and then adding the new column m. While one could do the same for $\mathbf{D}_F^k$, it is better to form a QR factorization of $\mathbf{D}_F^k$ and update that factorization as the iteration progresses [38, 190, 194, 201].

We will discuss the updating of $\mathbf{D}_F$ in a more abstract way. Suppose $\mathbf{A} = \mathbf{QR}$ is the QR factorization of an $N \times m$ matrix $\mathbf{A}$. If we wish to simply add a column to $\mathbf{A}$, we can update the QR factorization by orthogonalizing the new column against the columns of $\mathbf{Q}$ with, for example, the classical Gram–Schmidt algorithm applied twice. This is what we do in the **kl_gmres.jl** code from Chapter 3.

If we need to delete the first column of $\mathbf{A}$ and then add a new column, the method is more subtle. Let the QR factorization of $\mathbf{A}$ be

$$\mathbf{A} = (\mathbf{a}_1, \mathbf{a}_2, \ldots, \mathbf{a}_m) = \mathbf{QR},$$

where $\mathbf{Q}$ is $N \times m$ and $\mathbf{R}$ is $m \times m$. Now let $\tilde{\mathbf{R}}$ be the $m \times (m - 1)$ matrix formed by eliminating the first column of $\mathbf{R}$. It is easy to see that

$$\hat{\mathbf{A}} = (\mathbf{a}_2, \ldots, \mathbf{a}_m) = \mathbf{Q}\tilde{\mathbf{R}}.$$

Of course $\mathbf{Q}\tilde{\mathbf{R}}$ is not the QR factorization of $\hat{\mathbf{A}}$ because the matrices are the wrong size. However, if

$$\tilde{\mathbf{R}} = \mathbf{Q}^1 \hat{\mathbf{R}}$$

is the QR factorization of $\tilde{\mathbf{R}}$ and $\hat{\mathbf{Q}} = \mathbf{Q}\mathbf{Q}^1$ then $\hat{\mathbf{A}} = \hat{\mathbf{Q}}\hat{\mathbf{R}}$ is the QR factorization of $\hat{\mathbf{A}}$ and we are now prepared to use Gram–Schmidt orthogonalization to add the new column.

There are a few details one must address within an implementation of Anderson acceleration. Since $\tilde{\mathbf{R}}$ is an upper Hessenberg matrix, one could follow the implementation of GMRES [107, 170] and use Givens rotations to perform the factorization. While that would take fewer floating point operations that using the standard QR factorization from Julia, it makes very little difference since m is typically quite small in Anderson acceleration, so we simply use the **qr** command in Julia.

More importantly, when multiplying $\mathbf{Q}$ by $\mathbf{Q}^1$, one must store the intermediate result before overwriting the first $m - 1$ columns of $\mathbf{Q}$. This adds to the storage one needs for Anderson acceleration. Counting $\mathbf{D}_F$, $\mathbf{D}_G$, and the product $\mathbf{Q}\mathbf{Q}^1$ we have $3m - 1$ vectors. One can reduce this by performing the product $\mathbf{Q}\mathbf{Q}^1$ in blocks of rows, and that is an option in our solver **aasol.jl**. Using that option saves storage but is slower, and we only recommend it if storage is critical.

4.4 • Two Examples

4.4.1 • Chandrasekhar H-Equation

Our first example is the Chandrasekhar H-equation with $c = 0.99$ and initial iterate of $H(\mu) \equiv 1$. The choice of initial iterate is important for the discussion of the results. With this initial iterate, Picard iteration (Anderson(0) is also interesting) constructs the Taylor series of $H(\mu, c)$ as a function of the parameter c. The solution of interest is analytic in (complex!) c for $|c| < 1$. The kth Picard iteration is, in fact, a better approximation than the kth-order Taylor expansion. For this reason Picard iteration converges faster than Corollary 4.3 predicts.

We will discuss this a bit more in the context of the integral equation, taking the analysis from [33]. Everything we say also holds for the discrete problem. We use the notation from section 4.1.2. So

$$\mathcal{G}(H) = H - \mathcal{F}(H) \quad \text{and} \quad \mathbf{G}(\mathbf{x}) = \mathbf{x} - \mathbf{F}(\mathbf{x}),$$

where $\mathcal{F}$ is defined by (2.3). One can estimate the spectral radius of $\mathcal{G}'(H^*)$ analytically (and we do this in section 5.2),

$$\rho(\mathcal{G}'(H^*)) \leq 1 - \sqrt{1 - c}. \tag{4.20}$$

So, for $c = 0.99$, $\rho(\mathbf{G}'(H^*)) \leq 0.9$. We see from Figure 4.1 that the residual for Picard iteration has decreased by a factor of roughly 0.01 after 20 iterations. The spectral radius estimate (4.20) would predict a reduction of $0.9^{20} \approx 0.12$. In fact, if

$$H^*(\mu, c) = \sum_{m=0}^{\infty} c^m a_m(\mu)$$

is the Taylor expansion of H^* in c, then the coefficient functions $\{a_m(\mu)\}$ are nonnegative for $0 \leq \mu \leq 1$. Moreover the series converges for $c = 1$. Hence, if H_k is the kth Picard

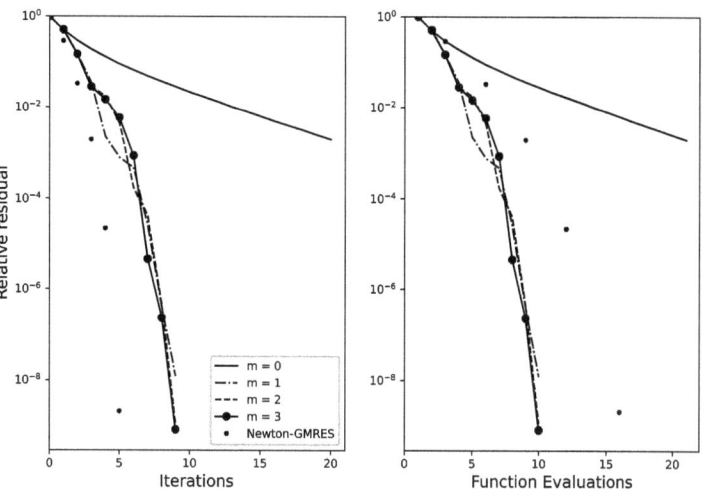

Figure 4.1: Anderson acceleration for the H-equation. $c = 0.99$.

iteration and $H_0 \equiv 1$, then for all $k \geq 0$ and $c, \mu \in [0, 1]$,

$$H_k(\mu, c) \leq H_{k+1}(\mu, c) \leq H^*(\mu, c).$$

We compare Newton-GMRES, the most efficient solver so far in this book, with Anderson(m) for $m = 0, 1, 2, 3$. While Newton-GMRES takes fewer nonlinear iterations, Anderson(m) for $m > 0$ takes significantly fewer function evaluations. Note that for this example there is little difference between $m = 1, 2, 3$, while Picard iteration failed to converge in 20 iterations.

4.4.2 ▪ Convection-Diffusion Equation

In this section we formulate the left-preconditioned nonlinear convection-diffusion equation as a fixed point problem. We let $\mathbf{F}$ be the nonlinear residual for the convection-diffusion problem (2.13) and let $\mathbf{P}$ be the fast solver for the high-order term. Following the discussion in section 3.3 we let

$$\mathbf{F}_P(\mathbf{x}) = \mathbf{PF}(\mathbf{x})$$

and then use (4.11) to obtain the fixed point map

$$\mathbf{G}(\mathbf{x}) = \mathbf{x} - \mathbf{PF}(\mathbf{x}).$$

We build the fixed point map **hardleftFix!** with the function **hardleft!** (see section 3.7.2) using (4.11). The Julia function **axpy!** is a call to the BLAS [7, 57]. The author of this book likes to make BLAS calls, which can reduce the allocation burden in Julia.

```
"""
hardleftFix!(FV, u, pdata)
Fixed point form of the left preconditioned nonlinear
convection-diffusion equation
"""
function hardleftFix!(FV, u, pdata)
FV = hardleft!(FV, u, pdata)
# G(u) = u - FV
axpby!(1.0, u, -1.0, FV)
return FV
end
```

While $\mathbf{G}$ is a nonlinear integral operator, the map is not a contraction and both Picard iteration and Anderson(1) diverge. (Try it yourself!) However, Anderson(m) for $m \geq 2$ does converge and, as Figure 4.2 illustrates, takes fewer function evaluations (but more nonlinear iterations) than Newton-GMRES for the left-preconditioned nonlinear problem (NewtonL) or a Newton-GMRES iteration with right preconditioning for the linear equation for the step (NewtonR). Unlike Newton-GMRES, which manages the residual norm decrease with a line search, Anderson(m) comes with no guarantee that the residual norms decrease as the iteration progresses, and they do not in this example (see iteration 3 for all m and the very irregular residual history for $m = 2$).

Remember that Anderson(m) needs roughly $3m$ vectors, while GMRES(m) needs roughly m vectors. For large N and m this can be a serious problem for Anderson(m).

It is also interesting to see that Anderson(10) **seems** to perform best, and hence one gains something (fewer function evaluations) for the increased storage burden. However (see section 4.5.3), function evaluations do not fully reflect the cost of the solve for this application. The reader should play with m and see what happens as m increases.

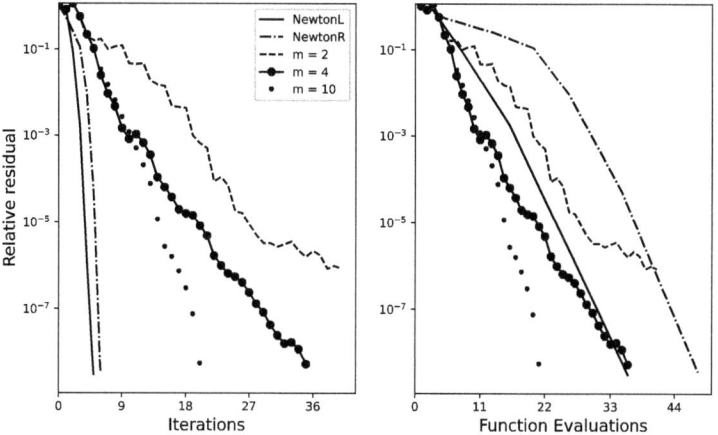

Figure 4.2: Anderson acceleration for the convection-diffusion equation.

4.5 ▪ Notebook: Solver for Chapter 4

4.5.1 ▪ aasol.jl

As usual, we begin with the docstrings. You should not expect the iteration statistics to agree to high precision across operating systems because the optimization problems are so ill-conditioned. This is especially the case with the example from [192].

```
[2]: ?aasol
```

```
[2]: aasol(GFix!, x0, m, Vstore; maxit=20, rtol=1.e-10, atol=1.e-10,
      beta=1.0, pdata=nothing, keepsolhist = false)

      C. T. Kelley, 2022

      Julia code for Anderson acceleration.  Nothing fancy.

      Solvers fixed point problems x = G(x).

      You must allocate storage for the function and fixed point map
      history -> in the calling program <- in the array Vstore.

      For an n dimensional problem with Anderson(m), Vstore must have
      at least 2m + 4 columns and 3m + 3 is better.  If m=0 (Picard)
      then V must have at least 4 columns.

      Inputs:
```

- GFix!: fixed-point map, the ! indicates that GFix! overwrites G, your preallocated storage for the function value G=G(xin).

  So G=GFix!(G,xin) or G=GFix!(G,xin,pdata) returns G=G(xin)

  Your GFix function MUST end with -> return G <-. See the example in the docstrings.

- x0: Initial iterate. It is a vector of size N

  You should store it as (N) and design G! to use vectors of size (N). If you use (N,1) consistently instead, the solvers may work, but I make no guarantees.

- m: depth for Anderson acceleration. m=0 is Picard iteration

- Vstore: Working storage array. For an n dimensional problem Vstore should have at least 3m+3 columns unless you are storage bound. If storage is a problem, then you can allocate a minimum of 2m+4 columns. The smaller allocation exacts a performance penalty, especially for small problems and small values of m. So for Anderson(3), Vstore should be no smaller than zeros(N,8) with zeros(N,11) a better choice. Vstore needs to allocate for the history of differences of the residuals and fixed point maps. The extra m-1 columns are for storing intermediate results in the downdating phase of the QR factorization for the coefficient matrix of the optimization problem. See the notebook or the print book for the details of this mess.

  If m=0, then Vstore needs 4 columns.

Keyword Arguments (kwargs):

maxit: default = 20

limit on nonlinear iterations

rtol and atol: default = 1.e-10

relative and absolute error tolerances

beta:

Anderson mixing parameter. Changes G(x) to (1-beta)x + beta G(x). Equivalent to accelerating damped Picard iteration. The history vector is the one for the damped fixed point map, not the original one. Keep this in mind when comparing results.

pdata:

precomputed data for the fixed point map. Things will go better if you use this rather than hide the data in global variables within the module for your function.

keepsolhist: default = false

Set this to true to get the history of the iteration in the
output tuple. This is on by default for scalar equations and
off for systems. Only turn it on if you have use for the data,
which can get REALLY LARGE.

Output:

  • A named tuple (solution, functionval, history, stats, idid,
    errcode, solhist)

where

- solution = converged result

- functionval = G(solution) You might want to use functionval
as your solution since it's a Picard iteration applied to the
converged Anderson result. If G is a contraction it will be
better than the solution.

- history = the vector of residual norms ($||x-G(x)||$) for the
iteration

- stats = named tuple (condhist, alphanorm) of the history of
the condition numbers of the optimization problem and l1 norm
of the coefficients. This is only for diagnosing problems
and research. Condihist[k] and alphanorm[k] are the condition
number and coefficient norm for the optimization problem that
computes iteration k+1 from iteration k.

I record this for iterations k=1, ... until the final iteration
K. So I do not record the stats for k=0 or the final iteration.
We did record the data for the final iteration in Toth/Kelley
2015 at the cost of an extra optimization problem solve. Since
we've already terminated, there's not any point in collecting
that data.

Bottom line: if history has length K+1 for iterations 0 ... K,
then condhist and alphanorm have length K-1.

- idid=true if the iteration succeeded and false if not.

- errcode = 0 if the iteration succeeded

    = -1 if the initial iterate satisfies the termination criteria

    = -2 if || residual || > div_test || residual_0 ||
        I have fixed div_test = 1.e4 for now. I terminate
        the iteration when this happens to avoid generating
        Infs and/or NaNs.

```
      = 10 if no convergence after maxit iterations
```

- solhist:

This is the entire history of the iteration if you've set keepsolhist=true

solhist is an N x K array where N is the length of x and K is the number of iterations + 1.

Examples for aasol

Duplicate Table 1 from Toth-Kelley 2015. The final entries in the condition number and coefficient norm statistics are never used in the computation and we don't compute them in Julia. See the docstrings, notebook, and the print book for the story on this.

```
julia> function tothk!(G, u)
          G[1]=cos(.5*(u[1]+u[2]))
          G[2]=G[1]+ 1.e-8 * sin(u[1]*u[1])
          return G
          end
tothk! (generic function with 1 method)

julia> u0=ones(2,); m=2; vdim=3*m+3; Vstore = zeros(2, vdim);
julia> aout = aasol(tothk!, u0, m, Vstore; rtol = 1.e-10);
julia> aout.history
8-element Vector{Float64}:
 6.50111e-01
 4.48661e-01
 2.61480e-02
 7.25389e-02
 1.53107e-04
 1.18513e-05
 1.82466e-08
 1.04725e-13


julia> [aout.stats.condhist aout.stats.alphanorm]
6×2 Matrix{Float64}:
 1.00000e+00   1.00000e+00
 2.01556e+10   4.61720e+00
 1.37776e+09   2.15749e+00
 3.61348e+10   1.18377e+00
 2.54948e+11   1.00000e+00
 3.67694e+10   1.00171e+00
```

Now we put a mixing or damping parameter in there with beta = .5. This example is nasty enough to make mixing do well. Keep in mind that the history is for the damped residual, not the original one.

```
julia> bout=aasol(tothk!, u0, m, Vstore; rtol = 1.e-10, beta=.5);

julia> bout.history
7-element Vector{Float64}:
 3.25055e-01
 3.70140e-02
 1.81111e-03
 9.55308e-04
 1.25936e-05
 1.40854e-09
 2.18196e-12
```

H-equation example with m=2. This takes more iterations than Newton, which should surprise no one.

```
julia> n=16; x0=ones(n,); Vstore=zeros(n,20); m=2;
julia> hdata=heqinit(x0,.99);
julia> hout=aasol(HeqFix!, x0, m, Vstore; pdata=hdata);
julia> hout.history
12-element Vector{Float64}:
 1.47613e+00
 7.47800e-01
 2.16609e-01
 4.32017e-02
 2.66867e-02
 6.82965e-03
 2.70779e-04
 6.51027e-05
 7.35581e-07
 1.85649e-09
 4.94803e-10
 5.18866e-12
```

4.5.2 ▪ The Mixing Parameter β

In this section we show how the mixing parameter can be used to make the fixed point map **hardleftFix!** (see section 4.4) for the convection-diffusion example contractive. You will see that the cost for that mixing is high.

We will use a mixing parameter $\beta = 0.2$ The code for this example is **mixing_pde.jl** in the **src/Chapter4** directory in the notebook repository. We will compare $m = 0, 2, 4, 10$ (see Figure 4.3). $m = 0$ (Picard) failed to converge for the undamped problem, but does converge, albeit slowly, here.

```
[3]: mixing_pde();
```

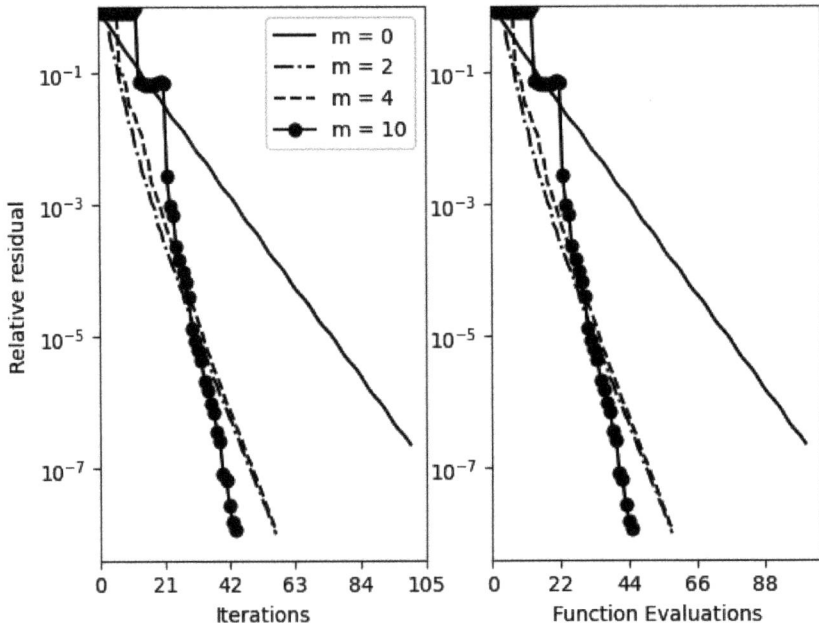

Figure 4.3: Anderson mixing for convection-diffusion problem.

```
Failure to converge after 100 iterations for m=0 in aasol.jl
```

While the mixing parameter of 0.2 makes the map $\mathbf{G}_\beta$ a contraction, the performance of the iteration is not as good as the unmixed iteration from **Figure 4.2**. You should be aware that the residual in the figure is the residual $\mathbf{F}_\beta(\mathbf{x}) = \mathbf{x} - \mathbf{G}_\beta(\mathbf{x})$, which is not the same as the residual reported in **Figure 4.2**. The residual histories still tell an accurate story. The message here is that mixing will not solve all your problems.

4.5.3 ▪ Benchmarking Anderson Acceleration

The examples in this chapter are integral equations. The H-equation is clearly an integral equation, and the left preconditioned convection diffusion equation is as well. These problems are good candidates for Anderson acceleration, as the title of [6] indicates. So, it should not be a complete surprise that Anderson acceleration performs very well for these problems.

We begin with the H-equation and compare the previous best solver, Newton-GMRES, with the Eisenstat–Walker forcing term with Anderson acceleration. The difference in timings is what one would expect from the difference in function evaluations one sees in **Figure 4.1**.

```
[4]:  n=4096; m=3;
      # Initialize. Set up the kwargs.
      FS=ones(n,); FPK=ones(n,10); Vstore=zeros(n,3*m+3);
      x0=ones(n,); c=.5; hdata = heqinit(x0, c);
      kbargsew=(atol = 1.e-10, rtol = 1.e-10, eta=.9,
```

```
      fixedeta=false, pdata=hdata);
aaargs=(atol = 1.e-10, rtol = 1.e-10,
    pdata=hdata, maxit=20);
# Run the solvers
println("Newton-GMRES, Eisenstat-Walker");
@btime nsoli(heqf!, $x0, $FS, $FPK; kbargsew...);
println("Anderson acceleration");
@btime aasol(HeqFix!, $x0, $m, $Vstore; aaargs...);
```

```
Newton-GMRES, Eisenstat-Walker
  1.776 ms (340 allocations: 501.19 KiB)
Anderson acceleration
  928.795 mus (228 allocations: 398.28 KiB)
```

Figure 4.2 indicates that $m = 10$ is the best choice for Anderson acceleration for the left-preconditioned convection-diffusion equation. We compare that with Newton-GMRES. The results are closer than for the H-equation with Newton-GMRES performing slightly better. One reason for this is that the initial iterate is poor and Newton-GMRES is better able to handle that than Anderson acceleration. Function evaluations are inexpensive for this example, so that is a less reliable indicator of performance than it is for the H-equation.

We also show that mixing makes the performance of Anderson acceleration worse.

```
[5]:  np=31; mp=10; N=np*np;
      # Initialize. Set up the kwargs.
      pdata=pdeinit(np); fdata=pdata.fdata;
      u0=zeros(N); FV=copy(u0); JV=zeros(N,20);
      Vstore=zeros(N,3*mp+3); atol=1.e-8; rtol=1.e-8;
      kbargs=(pdata=pdata, rtol=rtol, atol=atol,
          maxit=20, eta=.1, fixedeta=false);
      aaargs=(pdata=pdata, rtol=rtol, atol=atol, maxit=40);
      baargs=(pdata=pdata, rtol=rtol, atol=atol,
          maxit=100, beta=.2);
      println("Newton-GMRES, Eisenstat-Walker");
      @btime nsoli(hardleft!, $u0, $FV, $JV; kbargs...);
      println("Anderson acceleration");
      @btime aasol(hardleftFix!, $u0, $mp, $Vstore; aaargs...);
      println("Anderson acceleration with mixing");
      @btime aasol(hardleftFix!, $u0, $mp, $Vstore; baargs...);
```

```
Newton-GMRES, Eisenstat-Walker
  1.520 ms (855 allocations: 184.30 KiB)
Anderson acceleration
  1.367 ms (929 allocations: 457.41 KiB)
Anderson acceleration with mixing
  3.307 ms (2198 allocations: 1.05 MiB)
```

The reader should play with the parameters for both **nsoli** and **aasol** in these examples. Do changes in the parameters change anything?

4.6 ▪ Projects

4.6.1 ▪ More on Mixing

See how mixing affects the performance of Anderson acceleration for the H-equation. Begin with $\beta = 0.9$ and see if any value of β is better than no mixing at all.

4.6.2 ▪ Explaining Results

Can you explain why Anderson acceleration did so well in **Section 4.4.2**? Do the results from [153] apply to this case?

4.6.3 ▪ The EDIIS Algorithm (Not Easy!)

- (easy part) Write a brute-force Anderson acceleration code that uses the formulation in **Algorithm 4.1** and solves the optimization problem by solving the normal equations. How does the solver respond to the ill-conditioning in the examples from the text?

- The EDIIS [124, 33] method adds a nonnegativity constraint to the optimization problem for Anderson acceleration. Modify your brute-force code to use the EDIIS algorithm. The nasty part of this problem is to pick a solver for the optimization problem that can deal with the ill-conditioning. The obvious choice is the trust-region method from [36], which is in current versions of MATLAB, for example. A choice less sensitive to ill-conditioning is the method from [74], which was in older versions of MATLAB. Look at [33] for some discussion and an example with the H-equation. Repeat that example for the convection-diffusion problem.

Chapter 5

Case Studies

Files for This Chapter

- From the Package repository:

    - Anderson acceleration solver: **/src/Solvers/aasol.jl**

    - Newton–Krylov solver: **/src/Solvers/nsoli.jl**

    - H-equation: **/src/TestProblems/Systems/Hequation.jl**

    - Case studies: **/src/TestProblems/CaseStudies**

        * Heat transfer: **CR_Heat.jl**

        * Pseudoarclength of the H-equation: **heq_continue.jl**

        * Continuation code: **knl_continue.jl**

- From the Notebook repository: **/src/Chapter5**
  Julia codes that generate the figures and tables

5.1 ▪ Conductive-Radiative Heat Transfer

In this section we consider a multiphysics problem where conductive and radiative heat transfer are coupled. Our description of the problem is from [14, 15, 109]. These models have applications to the study of porous materials such as fibers, powders, and foams used in insulation at either very low or very high temperatures, heat treatment of ceramics, and the thermal properties of coated materials [133, 179, 200].

5.1.1 ▪ The Equations

We begin with the continuous formulation of the normalized and dimensionless equations for the problem [133, 179, 180, 181]. We will focus on one of the simpler problems in [181]. Our formulation is taken directly from [14].

We will discuss one way to discretize the problem in section 5.1.4. We refer the reader to [133, 179, 200] for the derivation of the equations.

The unknowns are the dimensionless intensity $\psi(x, \mu)$ of radiation at a point x in the direction having cosine μ with the positive x axis and the dimensionless temperature $\Theta(x)$ at x.

The radiative transport equation is

$$\mu \frac{\partial \psi}{\partial x}(x, \mu) + \psi(x, \mu) = \frac{c(x)}{2} \int_{-1}^{1} \psi(x, \mu') \, d\mu' + (1 - c(x))\Theta^4(x) \qquad (5.1)$$

for $x \in (0, \tau)$ with boundary conditions

$$\psi(0, \mu) = \Theta_l^4, \ \mu > 0, \qquad (5.2)$$

and

$$\psi(\tau, \mu) = \Theta_r^4, \ \mu < 0. \qquad (5.3)$$

In (5.1)

$$0 \le c(x) \le 1 \text{ for all } x \in [0, \tau]. \qquad (5.4)$$

We assume that c is continuous.

ψ is the **angular flux**. Define the **scalar flux**

$$f(x) = \frac{1}{2} \int_{-1}^{1} \psi(x, \mu') \, d\mu'. \qquad (5.5)$$

The temperature Θ satisfies the boundary value problem

$$\frac{\partial^2 \Theta}{\partial x^2} = Q(x), \ x \in [0, \tau], \ \Theta(0) = \Theta_l, \Theta(\tau) = \Theta_r, \qquad (5.6)$$

and couples to the radiative transport equation by

$$Q(x) = \frac{1}{2N_c} \frac{d}{dx} \int_{-1}^{1} \mu' \psi(x, \mu') \, d\mu'. \qquad (5.7)$$

In (5.7) N_c is the conduction to radiation parameter [133].

The order of integration and differentiation in (5.7) can be changed [109] and we obtain, using the transport equation (5.1) and the definition of f, (5.5),

$$\frac{d}{dx} \int_{-1}^{1} \mu' \psi(x, \mu') \, d\mu' = \int_{-1}^{1} \mu' \frac{\partial}{\partial x} \psi(x, \mu') \, d\mu'$$

$$= -2(1 - c(x))f(x) + 2(1 - c(x))\Theta^4(x). \qquad (5.8)$$

Hence,

$$Q(x) = \alpha(x)(\Theta^4(x) - f(x)), \ 0 < x < \tau, \qquad (5.9)$$

where

$$\alpha(x) = (1 - c(x))/N_c. \qquad (5.10)$$

5.1.2 ▪ Formulation as a Fixed Point Problem

We will formulate the equations as a fixed point problem for Θ (i.e., we **expose** Θ). We could have equally well used a fixed point problem for f. Decisions like this are common for multiphysics problems, and choosing which variables to expose can be a subtle problem.

In section 5.1.3 we will show that (5.1) is equivalent to a linear integral equation for f,

$$f - \mathcal{L}_1 f = \mathcal{L}_2(\Theta^4) + g, \tag{5.11}$$

where $\mathcal{L}_1$ and $\mathcal{L}_2$ are compact operators on $C[0, \tau]$ and $g \in C[0, \tau]$ depends on the boundary data. We can solve (5.11) efficiently with GMRES, for example, and that is what we will do for our computations.

The operators $\mathcal{L}_1$ and $\mathcal{L}_2$ are related by

$$\mathcal{L}_1 w = \mathcal{K}(cw) \quad \text{and} \quad \mathcal{L}_2 w = \mathcal{K}((1-c)w).$$

The operator $\mathcal{K}$ is defined for $u \in C[0, \tau]$ by

$$\mathcal{K}(u)(x) = \int_0^\tau k(x, y) u(y) \, dy, \tag{5.12}$$

where

$$k(x, y) = \frac{1}{2} \int_0^1 \exp(|x - y|/\mu) \frac{d\mu}{\mu}.$$

Our discretization will be based on the derivation of (5.11) and not directly on the integral equation itself. In this way one does not have to struggle with a quadrature rule for integrals against $k(x, y)$.

So, given Θ, one can recover f via

$$f = (I - \mathcal{L}_1)^{-1}(\mathcal{L}_2(\Theta^4) + g). \tag{5.13}$$

Let D_2^{-1} be the solution operator for

$$\frac{\partial^2 u}{\partial x^2} = h(x); \ u(0) = u(\tau) = 0, \tag{5.14}$$

and

$$\Theta_0(x) = (1 - x)\Theta_l + x\Theta_r.$$

Then (5.6) is equivalent to

$$\Theta = D_2^{-1} Q + \Theta_0. \tag{5.15}$$

Recall that the discretization of the second derivative operator is a tridiagonal matrix and hence D_2^{-1} can be applied efficiently.

Hence, we can combine (5.13) and (5.15) to obtain a fixed point problem for Θ:

$$\Theta = \mathcal{G}(\Theta) \equiv D_2^{-1} Q + \Theta_0$$

$$= D_2^{-1}(\alpha(\Theta^4 - f)) + \Theta_0 \tag{5.16}$$

$$= D_2^{-1}(\alpha(\Theta^4 - (I - \mathcal{L}_1)^{-1}(\mathcal{L}_2(\Theta^4) + g))) + \Theta_0.$$

We will compare Newton–Krylov methods with Anderson acceleration for this problem. Either approach needs an efficient evaluation of the discretization of the fixed point map $\mathcal{G}$. The steps in the evaluation of $\mathcal{G}$ are as follows:

1. Given Θ solve the transport equation (5.8) by solving the integral equation form (5.13) with an iterative method.

2. Use the solution f of the transport equation from step 1 to build $Q = \alpha(\Theta^4 - f)$.

3. Compute $\mathcal{G}(\Theta) = D_2^{-1}Q + \theta_0$ with a fast solver for (5.14).

We have covered step 3 previously and will discuss step 1 in sections 5.1.3 and 5.1.4

5.1.3 ▪ Derivation of (5.11)

We begin with a more general form of the transport equation

$$\mu\frac{\partial\psi}{\partial x}(x,\mu) + \psi(x,\mu) = \frac{c(x)}{2}\int_{-1}^{1}\psi(x,\mu')\,d\mu' + q(x) \tag{5.17}$$

for $x \in (0,\tau)$ with boundary conditions

$$\psi(0,\mu) = F_l(\mu),\ \mu > 0; \quad \psi(\tau,\mu) = F_r(\mu),\ \mu < 0. \tag{5.18}$$

In (5.17) and (5.18) the $\tau < \infty$, $c,q \in C([0,\tau])$, and F_l and F_r are given continuous real valued functions of μ. It is known [27] that the boundary value problem (5.17)–(5.18) has a unique solution if $0 \le c(x) \le 1$.

We begin with the continuous form of the source iteration map. Suppose the scalar flux f is known; then we can recover ψ by integration. For $\mu > 0$ we integrate (5.1) forward in x, obtaining

$$\psi(x,\mu) = \frac{1}{\mu}\int_{0}^{x}\exp(-(x-y)/\mu)(cf+q)(y)\,dy + \exp(-x/\mu)F_l(\mu),\ \mu > 0. \tag{5.19}$$

Similarly, for $\mu < 0$, we integrate backwards to obtain

$$\begin{aligned}
\psi(x,\mu) &= -\frac{1}{\mu}\int_{x}^{\tau}\exp(-(x-y)/\mu)(cf+q)(y)\,dy \\
&\quad + \exp((\tau-x)/\mu)F_r(\mu) \\
&= \frac{1}{|\mu|}\int_{x}^{\tau}\exp(-|x-y|/|\mu|)(cf+q)(y)\,dy \\
&\quad + \exp(-|\tau-x|/|\mu|)F_r(\mu).
\end{aligned} \tag{5.20}$$

The two computations of the angular flux via (5.19) and (5.20) are a **transport sweep**.

Integrating (5.19) and (5.20) in μ and adding the results leads to a linear fixed point equation for f,

$$f = \mathcal{S}(f,F_l,F_r,q) \equiv \mathcal{K}(cf) + \mathcal{K}(q) + g, \tag{5.21}$$

where $\mathcal{K}$ is given by (5.12) and

$$g(x) = \frac{1}{2}\int_0^1 \exp(-x/v)F_l(v)\,dv$$
$$+ \frac{1}{2}\int_0^1 \exp(-(\tau - x)/v)F_r(-v)\,dv$$
$$= \mathcal{S}(0, F_l, F_r, 0).$$

This completes the derivation of (5.11) with

$$q = (1 - c)\Theta^4, \quad F_l = \Theta_l^4, \quad \text{and} \quad F_r = \Theta_r^4.$$

The map $\mathcal{S}$ is the **source iteration map**.

The **source iteration** method for solving (5.21) is simply

$$f_{n+1} = \mathcal{S}(f_n, F_l, F_r, q),$$

which is Picard iteration for the linear integral equation form of the transport equation. However, source iteration does not need to compute $k(x, y)$ or the integral operator. Instead the action of the integral operator on f and q and the construction of g are done by solving two initial value problems to obtain (5.19) and (5.20).

One must formulate the problem (5.21) as a linear equation to use a Krylov method. The way to do this is to solve

$$f - \mathcal{L}_1 f = b,$$

where the operator-vector product for $\mathcal{L}_1$ can be expressed as

$$\mathcal{L}_1 f = \mathcal{S}(f, 0, 0, 0) \quad \text{and} \quad b = \mathcal{S}(0, F_l, F_r, q).$$

Hence an operator-vector product costs one transport sweep. Note that there is no convenient matrix representation of $\mathcal{L}_1$ and that matrix-free methods such as Krylov methods or source iteration are the only sensible way to solve the problem.

5.1.4 ▪ Discretization

We begin with a spatial mesh on $[0, \tau]$ with mesh width $h = \tau/(n + 1)$ and $x_j = (j - 1)h$. We will discretize the map D_2 with the usual central difference scheme and let $\mathbf{D}_2$ be that discretization.

For the transport equation we will use the discrete ordinates approach [108, 126, 152].

We construct our sequence of approximate source iteration maps with the diamond difference discretization. We use a variation on the notation of [126]. We let $\mathbf{f}$, $\mathbf{q}$, and $\mathbf{c}$ be the evaluations of f, q, and c at the spatial grid points.

$$f_i = f(x_i), \quad q_i = q(x_i), \quad \text{and} \quad c_i = c(x_i).$$

We let $\{\mu_j\}_{j=1}^{N_A}$ and $\{w_j\}_{i=1}^{N_A}$ be the nodes and weights for a quadrature rule in angle. The optimal quadrature for this application is a **double Gauss** rule, which is Gaussian quadratures for each of the intervals $[-1, 0)$ and $(0, 1]$ separately [152]. We use a double

20-point rule for this application and obtain the nodes and weights from the Julia package **QuadGK.jl** [100]. One could also use **FastGaussQuadrature.jl** [145] or copy the data from an old table [43], which is what we did in [108].

We solve the subproblem for f with GMRES and give a tolerance of $tol = 10^{-12}$ to the linear solver. Therefore we expect the nonlinear residual to be accurate to (roughly) twelve figures. In light of this we set the difference increment for the finite difference Jacobian-vector product to $dx = 10^{-5}$.

The diamond difference method first computes approximations ψ_i^j to $\psi(x_i, \mu_j)$ by computing ψ_i^j for each i with j fixed. The discretized forward integration ($\mu_j > 0$) is

$$\mu_j \frac{\psi_i^j - \psi_{i-1}^j}{h} + \frac{\psi_i^j + \psi_{i-1}^j}{2}$$

$$= \frac{c_i f_i + c_{i-1} f_{i-1}}{2} + \frac{q_i + q_{i-1}}{2}, \, \mu_j > 0, \tag{5.22}$$

with initial data $\psi_1^j = \psi(0, \mu_j) = F_l(\mu_j)$. Similarly the discretized backward integration ($\mu_j < 0$) is

$$\mu_j \frac{\psi_i^j - \psi_{i+1}^j}{h} + \frac{\psi_i^j + \psi_{i+1}^j}{2}$$

$$= \frac{c_i f_i + c_{i+1} f_{i+1}}{2} + \frac{q_i + q_{i+1}}{2}, \, \mu_j < 0, \tag{5.23}$$

with final data $\psi_N^j = \psi(\tau, \mu_j) = F_r(\mu_j)$.

Following the forward and backward integration we compute the discrete source iteration map

$$\mathbf{S}(\mathbf{f})_i = \frac{1}{2} \sum_{j=1}^{N_A} \psi_i^j w_j. \tag{5.24}$$

Similarly to the derivation of the continuous source iteration map, $\mathbf{q}$ and the boundary data are built into $\{\psi_i^j\}$ by the way in which the integration is done. Hence we also write the discrete source iteration map as

$$\mathbf{S}(\mathbf{f}, F_l, F_r, \mathbf{q}).$$

So the discretized linear integral equation form of the transport equation is

$$\mathbf{f} = \mathbf{S}(\mathbf{f}, F_l, F_r, \mathbf{q}) \equiv \mathbf{L}_1(\mathbf{f}) + \mathbf{b},$$

where

$$\mathbf{L}_1 f = \mathbf{S}(f, 0, 0, 0) \quad \text{and} \quad \mathbf{b} = \mathbf{S}(0, F_l, F_r, \mathbf{q}).$$

We obtain the discretization of the linear map from Θ to $\mathbf{f}$ as before with the substitution $\mathbf{q} = (1 - c)\Theta^4$ where products are understood to be componentwise. We let $\mathbf{T}$ and $\mathbf{T}_0$ be the discretizations of Θ and Θ_0. The discretization of $\mathcal{L}_2$ is

$$\mathbf{L}_2(\mathbf{T}^4) = \mathbf{S}(0, 0, 0, (1 - c)\mathbf{T}^4).$$

The discretized fixed point problem for Θ is

$$\mathbf{T} = \mathbf{G}(\mathbf{T}) \equiv \mathbf{D}_2^{-1}(\alpha \mathbf{T}^4 - (I - \mathbf{L}_1)^{-1}(\mathbf{L}_2(\mathbf{T}^4) + \mathbf{g})) + \mathbf{T}_0. \qquad (5.25)$$

Here

$$\mathbf{g} = \mathbf{S}(0, F_l, F_r, 0).$$

The Julia code for the fixed point map is in **CR_Heat.jl** in the **/src/TestProblems/Case Studies/** directory.

5.1.5 ▪ Examples for Conductive-Radiative Heat Transfer

The example in this section is taken from [181]. Transport theory has a generous supply of benchmark problems and this is one. We begin by comparing results with the tables in [181]. We use an $N = 1001$ point grid so that the tabulated data are at mesh points. We consider the two problems given in Table 5.1.

We solve the problems with Anderson(2).

The results in Table 5.2 agree to all figures reported in column two of Tables 2 and 3 in [181].

We close this section with a comparison of Anderson(m) for several values of m and Newton-GMRES on three instances of the convective-radiative heat transfer problem. We begin with Problem 1 from Table 5.1 and then vary ω, τ, and θ_r. The first problem is easy, and Figure 5.1 shows that Anderson(2) is clearly a good choice.

Table 5.1: Two problems.

Problem 1:	$N_c = 0.05, \omega = 0.9, \tau = 1, \Theta_l = 1, \Theta_r = 0$
Problem 2:	$N_c = 0.05, \omega = 0.9, \tau = 1, \Theta_l = 1, \Theta_r = 0.5$

Table 5.2: Temperature for the two problems.

x	Θ_1	Θ_2
0.00	1.00000e+00	1.00000e+00
0.10	9.18027e-01	9.54270e-01
0.20	8.36956e-01	9.11008e-01
0.30	7.53557e-01	8.68433e-01
0.40	6.65558e-01	8.25127e-01
0.50	5.71475e-01	7.79940e-01
0.60	4.70505e-01	7.31936e-01
0.70	3.62437e-01	6.80375e-01
0.80	2.47544e-01	6.24709e-01
0.90	1.26449e-01	5.64610e-01
1.00	0.00000e+00	5.00000e-01

The fixed point maps are not contractive for the second two problems. For the problem in Figure 5.2 Anderson(m) shows irregular convergence, but for $m > 0$ recovers well. Figure 5.3 shows that Picard iteration diverges for the last of the problems. Our code **aasol.jl** halts the iteration with a failure message once the residual norm increases by a factor of 10^4, as it did for Picard iteration in this example. Notice that Newton-GMRES has a far more regular and predictable convergence behavior for the two noncontractive problems. This regularity is an important advantage, and it is not clear that Anderson acceleration is a reasonable alternative. For example, in both Figures 5.2 and 5.3 Anderson(10), needs over twice the storage of the Newton-GMRES iteration.

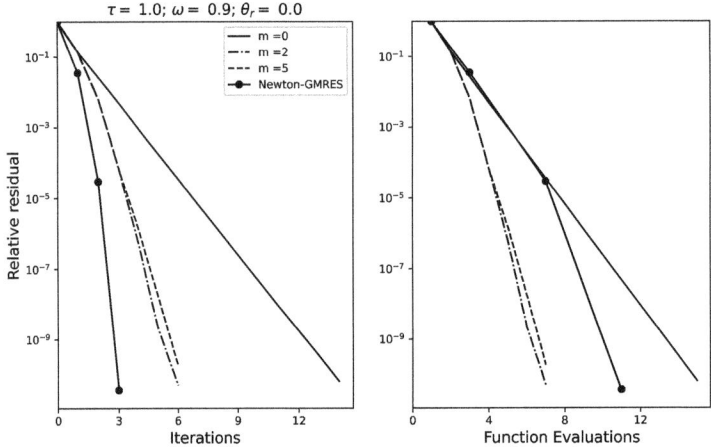

Figure 5.1: Easy heat transfer problem.

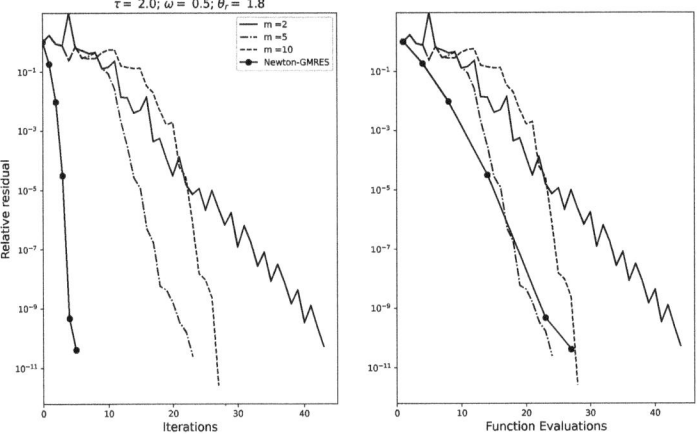

Figure 5.2: Less easy heat transfer problem.

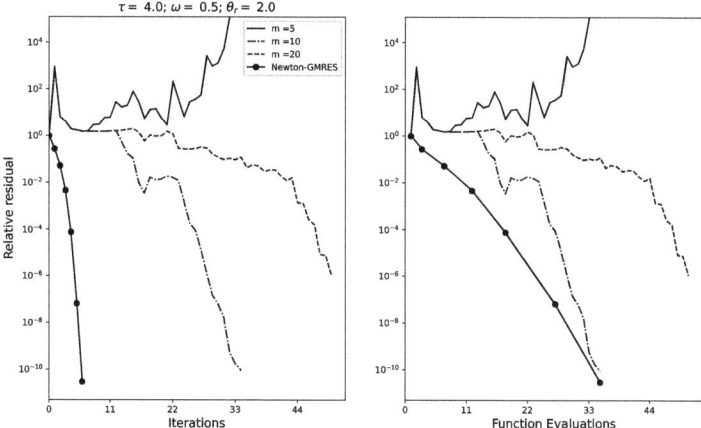

Figure 5.3: Hard heat transfer problem.

5.2 ▪ A Continuation Problem for the H-Equation

In this section we will explore some properties of the solution of H-equation as a function of the parameter c. As before, we will do the analysis for the continuous problem, but the results are valid for any discretization with a quadrature rule with positive weights that integrates constants exactly.

We have seen other examples of parameter-dependent nonlinear equations in the buckling beam problem from sections 2.7.6 and 3.5 and the cubic polynomial example in section 1.7.1. These examples have pitchfork bifurcations [106], and most path-following codes [56, 76, 125, 165, 171, 196, 198] have machinery to detect intersecting solution arcs and follow the different branches. Those methods are beyond the scope of this book. The Julia package **BifurcationKit.jl** [198] is quite complete. Our example of the H-equation is the simplest possible case.

5.2.1 ▪ Properties of H as a Function of c

Recall that the equation in $C[0, 1]$ is

$$\mathcal{F}(H)(\mu) = H(\mu) - \frac{1}{1 - \frac{c}{2} \int_0^1 \frac{\mu H(\mu)}{\mu + \nu} \, d\nu} = 0.$$

We will also refer to the fixed point form of the equation

$$H(\mu) = \mathcal{G}(H)(\mu) = \left(1 - \frac{c}{2} \int_0^1 \frac{\mu H(\mu)}{\mu + \nu} \, d\nu\right)^{-1}.$$

We will use an alternative form of the H-equation to compute the L^1 norm of H.

$$H(\mu) = 1 + \frac{c}{2} H(\mu) \int_0^1 \frac{\mu}{\mu + \nu} H(\nu) \, d\nu. \tag{5.26}$$

Integrating both sides of (5.26) yields

$$\|H\|_1 = 1 + \frac{c}{2} \int_0^1 \int_0^1 H(\mu)H(\nu)\frac{\mu}{\mu+\nu}\,d\nu.$$

Since μ and ν are variables of integration, we may interchange them to obtain

$$\|H\|_1 = 1 + \frac{c}{2} \int_0^1 \int_0^1 H(\mu)H(\nu)\frac{\nu}{\mu+\nu}\,d\nu.$$

When we average the two equations we obtain

$$\|H\|_1 = 1 + \frac{c}{4}\|H\|_1^2.$$

So

$$\|H\|_1 = \frac{1 \pm \sqrt{1-c}}{c/2}. \tag{5.27}$$

As we said in Chapter 2, the H-equation has two solutions for $0 < c < 1$. The **lower branch**, where $\|H\|_1 = \frac{1-\sqrt{1-c}}{c/2}$, is the solution of physical interest and the one you're likely to find with any iteration. In this section we will explore computation of the **upper branch**, where $\|H\|_1 = \frac{1+\sqrt{1-c}}{c/2}$.

Equation (5.27) implies that the H-equation has no real solutions for $c > 1$. This implies that the Fréchet derivative $\mathcal{F}'$ of $\mathcal{F}$ must be singular. The solution path cannot stop abruptly [40], so there must be an upper branch. Equation (5.27) implies that $\|H\|_1 \to \infty$ as $c \to 0$ on the upper branch.

We will briefly consider the details of the singularity before following the path. We begin by showing that $\mathcal{F}'(H)$ is nonsingular for $0 \le c < 1$ on the lower branch. To simplify the analysis we define the linear operator $\mathcal{L}$ on $C[0,1]$ by

$$(\mathcal{L}u)(\mu) = \int_0^1 \frac{\mu u(\nu)}{\mu+\nu}\,d\nu$$

and express the H-equation as

$$H = \frac{1}{1 - (c/2)\mathcal{L}H},$$

where the division is understood to be pointwise. Now note that, for $u \in C[0,1]$,

$$\mathcal{F}'(H)u = u - \frac{(c/2)\mathcal{L}u}{(1-(c/2)\mathcal{L}H)^2}$$

$$= u - (c/2)H^2\mathcal{L}u. \tag{5.28}$$

Here we use, as we have before, the simple trick to compute a Fréchet derivative,

$$\mathcal{F}'(H)u = \frac{d}{d\epsilon}\mathcal{F}(H + \epsilon u)\Big|_{\epsilon=0}.$$

Since $\mathcal{F}'$ is the sum of a compact operator and the identity, it is singular only if 0 is an eigenvector. So singularity of $\mathcal{F}'$ is equivalent to $\mathcal{G}'(H)$ having $\sigma(c) = 1$ as an eigenvalue.

When $c = 1$, the Perron theorem [104] and the positivity of the H-function imply that the largest eigenvalue in absolute value of $\mathcal{G}'(H)$ is positive and the corresponding eigenfunction does not change sign, and hence can be taken as nonnegative. That eigenvalue is $\sigma(1) = 1$ and the eigenfunction is $u(\mu) = \mu H(\mu)$. To see this use (5.27) and compute, with $\|H\|_1 = 2$,

$$\mathcal{G}'(H)(u)(\mu) = H^2(\mu)(1/2) \int_0^1 \frac{\mu v H(v)}{\mu + v}\, dv$$

$$= (\mu H(\mu)) H(\mu)(1/2) \int_0^1 \frac{v H(v)}{\mu + v}\, dv$$

$$= u(\mu) H(\mu)(1/2) \int_0^1 H(v) \left(1 - \frac{\mu}{\mu + v}\right) dv$$

$$= u(\mu) \left(H(\mu) - (1/2)(H\mathcal{L}H)(\mu)\right) = u(\mu).$$

Since $u \geq 0$, $\sigma(1) = 1$ is the Perron eigenvalue. Therefore the eigenvalue has multiplicity one.

The Perron theory is also applicable if $0 < c < 1$. Let $\sigma(c) > 0$ be the Perron eigenvalue of $\mathcal{G}'(H)$ with eigenfunction u. Set $u(\mu) = \mu H(\mu) p(\mu)$. Then

$$\sigma(c) p(\mu) = H(\mu)(c/2) \int_0^1 \frac{H(v) p(v) v}{\mu + v}\, dv \leq \|p\|_\infty H(\mu)(c/2) \int_0^1 \frac{H(v) v}{\mu + v}\, dv$$

$$\leq \|p\|_\infty H(\mu)(c/2) \int_0^1 H(v) \left(1 - \frac{\mu}{\mu + v}\right) dv$$

$$= \|p\|_\infty H(\mu)(1 - \sqrt{1 - c}) - (H(\mu) - 1)$$

$$= \|p\|_\infty (1 - H(\mu)\sqrt{1 - c}) \leq \|p\|_\infty (1 - \sqrt{1 - c}).$$

$$(5.29)$$

Hence, taking the L^∞ norm of the left side of (5.29), we have

$$\sigma(c) \leq 1 - \sqrt{1 - c} < 1.$$

Hence $\mathcal{F}'(H)$ is nonsingular for $0 < c < 1$ on the lower branch. This also proves (4.20) from section 4.4.1.

For $c = 1$, $\mathcal{F}'(H)$ is singular with a one-dimensional null space. To complete the analysis of the singularity we need to make the dependence in c explicit and write the equation as

$$\mathcal{F}(H, c) = 0.$$

We let $\mathcal{F}'_H$ be the Fréchet derivative in H and $\mathcal{F}'_c(H, c)$ be the derivative in c. At a solution, using (5.26),

$$\mathcal{F}'_c(H, c) = -H^2 \frac{1}{2} \mathcal{L}H,$$

and so, at $c = 1$,

$$\mathcal{F}'_c(H, 1) = -H(H - 1).$$

The singularity at $c = 1$ is a **simple fold** [106] . This means that $\mathcal{F}'(H)$ has a null space of dimension 1 and that $\mathcal{F}'_c(H, 1) \notin \mathcal{R}(\mathcal{F}'_H(H))$. Here $\mathcal{R}$ denotes the range of an operator.

In this case

$$\mathcal{R}(\mathcal{F}_H'(H)) = \left\{ u \in C[0,1] \mid \int_0^1 u(\nu)\psi(\nu)\,d\nu = 0 \right\},$$

where ψ is the eigenfunction of $\mathcal{F}_H'(H)^*$, the adjoint of $\mathcal{F}_H'(H)$ corresponding to the zero eigenvalue. The reader can verify that, for $c = 1$,

$$\psi(\mu) = (\mathcal{G}'(H)^*\psi)(\nu) = \frac{1}{2}\int_0^1 \frac{\nu}{\nu+\mu}H^2(\nu)\psi(\nu)\,d\nu$$

is satisfied by $\psi = H^{-1}$. Since $H \geq 1$ and $\mathcal{F}_c'(H,1)$ does not change sign, $\mathcal{F}_c'(H,1)$ cannot be in the range of $\mathcal{F}_H'(H)$. We will show why this fact is important in the next section.

5.2.2 ▪ Parameter Continuation

The object of continuation methods is to solve a parameter-dependent set of nonlinear equations

$$\mathbf{F}(\mathbf{x}, \lambda) = 0.$$

The output is a **solution arc** or **solution path**

$$\{\mathbf{x}(\lambda) \mid \mathbf{F}(\mathbf{x}(\lambda), \lambda) = 0, \lambda_0 \leq \lambda \leq \lambda_{max}\}.$$

The solution arc for the H-equation is a simple arc with no bifurcations. Because of this we can use a straightforward path following algorithm. The objective of this section is to show how a nonlinear solver can be modified in a simple way to resolve singularities such as the one for the H-equation at $c = 1$.

One simple approach to computing a solution path would be to identify a solution $\mathbf{x}(\lambda_0)$ for a particular value of λ and then increment λ by a small amount and use Newton's method with an initial iterate from the previous solution to compute the solution for the next step. In the case of the H-equation, where the parameter is c, one can use $c_0 = 0$ and $H \equiv 1$ to begin the continuation.

A candidate for the algorithm is **natural parameter continuation** where we increment the original parameter in the equation. Note that in the discussion in this section, as is common practice, the parameter is called λ in the context of a general method, but the actual name of the parameter (c in the case of the H-equation) is used when talking about specific examples.

ALGORITHM 5.1.
ParamCont$(\mathbf{F}, \mathbf{x}_0, \lambda_0, d_\lambda, \lambda_{max})$
    Solve $\mathbf{F}(\mathbf{x}, \lambda_0) = 0$ for $\mathbf{x} = \mathbf{x}(\lambda_0)$ with initial iterate $\mathbf{x}_0$.
    $\lambda = \lambda_0$
    while $\lambda \leq \lambda_{max}$ **do**
        Solve $\mathbf{F}(\mathbf{x}, \lambda + d_\lambda) = 0$ for $\mathbf{x} = \mathbf{x}(\lambda + d_\lambda)$ with initial iterate $\mathbf{x}(\lambda)$.
        $\lambda \leftarrow \lambda + d_\lambda$
    end while

One can solve the nonlinear equation for $\mathbf{x}$ with any iterative method. In this section we use only Newton-GMRES with a finite difference Jacobian-vector product. Our reasons for this are that we can use the same method for all continuation algorithms and that it is the most efficient choice for the H-equation. We will discuss this in more depth later.

One other point is that $\mathbf{x}(\lambda)$ is not the best choice for the initial iterate (the predictor) for the nonlinear equation for $\mathbf{x}(\lambda + d_\lambda)$. We use the linear (secant) predictor [106] in our codes,

$$\mathbf{x}_0(\lambda + d_\lambda) = 2\mathbf{x}(\lambda) - \mathbf{x}(\lambda - d_\lambda),$$

once we have two solutions on the path.

Natural parameter continuation is rarely used in practice because it is unable to resolve even the simplest singularities. The reason for this is clear for the H-equation. If we set $c_0 = 0$ and start the continuation, the iteration has to fail as soon as $c + d_c > 1$ because there are no real solutions for $c > 1$.

The singularity of the H-equation is particularly simple. Following [106] we consider a solution arc

$$\mathcal{S} = \{\mathbf{x}(\lambda) \mid \mathbf{F}(\mathbf{x}(\lambda), \lambda) = 0, \lambda_0 \leq \lambda \leq \lambda_{max}\}.$$

We say the path is **regular** if

$$Rank[\mathbf{F}'_\mathbf{x}, \mathbf{F}'_\lambda] = N \tag{5.30}$$

for all points on the path. Here, as in the previous section, $\mathbf{F}'_\mathbf{x}$ is the Jacobian of $\mathbf{F}$ in the $\mathbf{x}$ variables and $\mathbf{F}'_\lambda = d\mathbf{F}/d\lambda$.

The solution path for the H-equations is regular as we can see from the following lemma (Lemma 4.2 from [106]).

Lemma 5.1. $Rank[\mathbf{F}'_\mathbf{x}, \mathbf{F}'_\lambda] = N$ if and only if one of the following applies:

- $\mathbf{F}'_\mathbf{x}$ is nonsingular.

- The null space of $\mathbf{F}'_\mathbf{x}$ has dimension 1 and $\mathbf{F}'_\lambda \notin \mathcal{R}(\mathbf{F}'_\mathbf{x})$.

We showed that (at least for the lower branch up to $c = 1$) the solution path for the H-equation is regular. The upper branch is also regular with $\mathcal{F}'(H)_H$ nonsingular for the upper branch where $0 < c < 1$. We leave it to the reader to verify that.

Not all solution arcs are regular. The pitchfork bifurcation where two solution arcs intersect is not a regular point, and the algorithm we use in this section is not sufficient to identify a such a bifurcation and differentiate between stable and unstable branches.

One can resolve any singularity on a regular path with an appropriate reparameterization. One clever way is to swap λ with a component of $\mathbf{x}$ (see [165]). In the case of the discretized H-equation, making c one of the unknowns and $H(1)$ the parameter will work fine and there will be no singularity. However, this approach is highly problem dependent.

5.2.3 ▪ Pseudo-arclength Continuation

We will compute regular paths with **pseudo-arclength continuation** [106]. Here we introduce a parameter s for the path, so

$$\mathcal{S} = \{(\mathbf{x}(s), \lambda(s)) \mid 0 \le s \le s_{max}\},$$

and we apply the simple continuation algorithm **ParamCont** to the path as a function of s. In order to do this we need to augment $\mathbf{F}$ with a new equation for λ.

So, given $\mathbf{x}(s)$ and $\lambda(s)$, we solve

$$\hat{\mathbf{F}}(\mathbf{x}, \lambda, s) = \begin{pmatrix} \mathbf{F}(\mathbf{x}, \lambda) \\ N(\mathbf{x}, \lambda, s) \end{pmatrix} \tag{5.31}$$

for $\mathbf{x}(s+ds)$ and $\lambda(s+ds)$.

Here the **normalization equation**

$$N(\mathbf{x}, \lambda, s) \equiv \nu \dot{\mathbf{x}}(s)^T (\mathbf{x} - \mathbf{x}(s)) + \dot{\lambda}(s)(\lambda - \lambda(s)) - ds = 0 \tag{5.32}$$

is intended to let the artificial parameter s play the role of arclength [106].

In (5.32) $\dot{\mathbf{x}}$ and $\dot{\lambda}$ are approximations to $d\mathbf{x}/ds$ and $d\lambda/ds$. The approximations need not be highly accurate, only good enough to make sure that the continuation moves in the correct direction along the path and does not reverse itself. After we have two points on the path, we use

$$\dot{\mathbf{x}}(s) = \frac{\mathbf{x}(s) - \mathbf{x}(s - ds)}{ds} \quad \text{and} \quad \dot{\lambda}(s) = \frac{\lambda(s) - \lambda(s - ds)}{ds}.$$

The parameter ν in (5.32) is a scaling parameter designed to make λ and $\mathbf{x}$ equally important in the continuation [171]. For the H-equation, we use $\nu = 100/N$ so that the scalar product is the product of the composite midpoint approximation of the L^2 inner product for functions with a weighting factor that makes the continuation perform better.

If the path is regular, then $\hat{\mathbf{F}}'$ is nonsingular. The Jacobian is

$$\hat{\mathbf{F}}' = \begin{pmatrix} \mathbf{F}'_{\mathbf{x}} & \mathbf{F}'_{\lambda} \\ \dot{\mathbf{x}}^T & \dot{\lambda} \end{pmatrix}.$$

Regularity implies that the first N rows of the $(N+1) \times (N+1)$ matrix $\hat{\mathbf{F}}'$ have full rank (N). The last row is nonzero on a regular path, and this (see Lemma 4.9 in [106]) implies that $\hat{\mathbf{F}}'$ is nonsingular.

The last subtle point is how one solves the linear equation for the Newton step in $(\mathbf{x}, \lambda)$. We will exploit the integral equation structure of the H-equation and use **nsoli.jl** with Newton-GMRES for this. One could also use a direct method and exploit any structure in $\mathbf{F}'$, but that is a bit more difficult [106].

We apply these ideas to continue the solution of the H-equation in c. Figure 5.4 is the result. The horizontal axis is c and the vertical axis is the midpoint rule approximation of the L^1 norm of H. We used a constant value of $ds = 1/100$, which worked fine until the predictor failed to produce an initial iterate from which **nsoli.jl** would converge within 20 iterations. While reducing ds or using a higher-order predictor would enable one to move further along the path, the fact that $\|H\|_1 \to \infty$ as $c \to 0$ on the upper branch implies that you won't get much farther than the results from the figure.

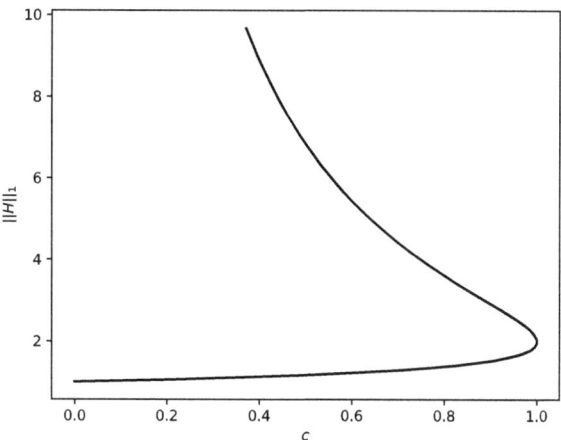

Figure 5.4: Continuation for the H-equation.

5.3 ▪ Projects

5.3.1 ▪ More on Heat Transfer

There are more interesting examples of conductive-radiative heat transfer in [181]. Extend the methods from this chapter to the more general boundary conditions in equations (10a)–(10b) on page 274 of [181].

5.3.2 ▪ Advanced Continuation

Explore solving the H-equation and the buckling beam equation with a more sophisticated continuation code than the simple one we used in this section. The Julia package **BifurcationKit.jl** [198] is a good place to start.

Bibliography

[1] N. J. ADAMOWICZ, *Perturbed dynamic mode decomposition acceleration of source iteration for fixed-source neutron transport problems*, in Proceedings of International Conference on Mathematics and Computational Methods Applied to Nuclear Science & Engineering, 2021, American Nuclear Society, La Grange Park, IL, 2021, pp. 1408–1417. (Cited on p. 134)

[2] E. L. ALLGOWER, K. BÖHMER, F. A. POTRA, AND W. C. RHEINBOLDT, *A mesh-independence principle for operator equations and their discretizations*, SIAM J. Numer. Anal., 23 (1986), pp. 160–169. (Cited on pp. 59, 128)

[3] A. AMERIO, *From zero to Julia!* 2020, https://techytok.com/from-zero-to-julia. (Cited on p. xviii)

[4] H. AN, X. JIA, AND H. F. WALKER, *Anderson acceleration and application to the three-temperature energy equations*, J. Comp. Phys., 347 (2017), pp. 1–19, https://doi.org/10.1016/j.jcp.2017.06.031. (Cited on pp. 134, 135)

[5] R. ANANTHARAMAN, *AlgebraicMultigrid.jl*, Julia Package, 2021, https://github.com/JuliaLinearAlgebra/AlgebraicMultigrid.jl. (Cited on p. 100)

[6] D. G. ANDERSON, *Iterative procedures for nonlinear integral equations*, J. ACM, 12 (1965), pp. 547–560, https://doi.org/10.1145/321296.321305. (Cited on pp. 22, 133, 146)

[7] E. ANDERSON, Z. BAI, C. BISCHOF, J. DEMMEL, J. DONGARRA, J. DU CROZ, A. GREENBAUM, S. HAMMARLING, A. MCKENNEY, S. OSTROUCHOV, AND D. SORENSEN, *LAPACK Users' Guide*, 2nd ed., SIAM, Philadelphia, 1992. (Cited on pp. 28, 50, 51, 56, 140)

[8] D. Y. ANISTRATOV, J. M. COALE, J. S. WARSA, AND J. H. CHUNG, *Multilevel second-moment methods with group decomposition for multigroup transport problems*, in Proceedings of International Conference on Mathematics and Computational Methods Applied to Nuclear Science & Engineering, 2021, American Nuclear Society, La Grange Park, IL, 2021, pp. 404–413. (Cited on p. 134)

[9] L. ARMIJO, *Minimization of functions having Lipschitz-continuous first partial derivatives*, Pacific J. Math., 16 (1966), pp. 1–3. (Cited on p. 12)

[10] U. M. ASCHER AND L. R. PETZOLD, *Computer Methods for Ordinary Differential Equations and Differential Algebraic Equations*, SIAM, Philadelphia, 1998. (Cited on pp. 53, 61, 66)

[11] K. E. ATKINSON, *An Introduction to Numerical Analysis*, 2nd ed., John Wiley and Sons, New York, 1989. (Cited on p. xii)

[12] S. BALAY, S. ABHYANKAR, M. F. ADAMS, J. BROWN, P. BRUNE, K. BUSCHELMAN, L. DALCIN, A. DENER, V. EIJKHOUT, W. D. GROPP, D. KARPEYEV, D. KAUSHIK, M. G. KNEPLEY, D. A. MAY, L. C. MCINNES, R. T. MILLS, T. MUNSON, K. RUPP, P. SANAN, B. F. SMITH, S. ZAMPINI, H. ZHANG, AND H. ZHANG, *PETSc Web page* 2019, https://www.mcs.anl.gov/petsc. (Cited on pp. xii, 21)

[13] S. BALAY, S. ABHYANKAR, M. F. ADAMS, J. BROWN, P. BRUNE, K. BUSCHELMAN, L. DALCIN, A. DENER, V. EIJKHOUT, W. D. GROPP, D. KARPEYEV, D. KAUSHIK, M. G. KNEPLEY, D. A. MAY, L. C. MCINNES, R. T. MILLS, T. MUNSON, K. RUPP, P. SANAN, B. F. SMITH, S. ZAMPINI, H. ZHANG, AND H. ZHANG, *PETSc users manual*, Tech. Report ANL-95/11 - Revision 3.13, Argonne National Laboratory, 2020, https://www.mcs.anl.gov/petsc. (Cited on p. 21)

[14] J. M. BANOCZI AND C. T. KELLEY, *A fast multilevel algorithm for the solution of nonlinear systems of conductive-radiative heat transfer equations*, SIAM J. Sci. Comp., 19 (1998), pp. 266–279. (Cited on p. 149)

[15] J. M. BANOCZI AND C. T. KELLEY, *A fast multilevel algorithm for the solution of nonlinear systems of conductive-radiative heat transfer equations in two space dimensions*, SIAM J. Sci. Comp., 20 (1999), pp. 1214–1228. (Cited on p. 149)

[16] D. J. BATES, J. D. HAUENSTEIN, A. J. SOMMESE, AND C. W. WAMPLER, *Numerically Solving Polynomial Systems with Bertini*, SIAM, Philadelphia, 2013. (Cited on pp. 15, 25)

[17] J. BEZANSON, A. EDELMAN, S. KARPINSKI, AND V. B. SHAH, *Julia: A fresh approach to numerical computing*, SIAM Review, 59 (2017), pp. 65–98. (Cited on pp. xi, xviii)

[18] W. BIAN, X. CHEN, AND C. T. KELLEY, *Anderson acceleration for a class of nonsmooth fixed-point problems*, SIAM J. Sci. Comp., 43 (2021), pp. S1–S20, https://doi.org/10.1137/20M132938X. (Cited on p. 136)

[19] P. BREIDING AND S. TIMME, *HomotopyContinuation.jl: A package for homotopy continuation in Julia*, in International Congress on Mathematical Software, Springer, Cham, 2018, pp. 458–465. (Cited on pp. 25, 26)

[20] K. E. BRENAN, S. L. CAMPBELL, AND L. R. PETZOLD, *Numerical Solution of Initial Value Problems in Differential-Algebraic Equations*, no. 14 in Classics in Applied Mathematics, SIAM, Philadelphia, 1996. (Cited on pp. 11, 53, 66)

[21] R. P. BRENT, *Some efficient algorithms for solving systems of nonlinear equations*, SIAM J. Numer. Anal., 10 (1973), pp. 327–344. (Cited on pp. 56, 92)

[22] E. L. BRIGGS, D. J. SULLIVAN, AND J. BERNHOLC, *Large-scale electronic-structure calculations with multigrid acceleration*, Phys. Rev. B, 52 (1995), pp. R5471–R5474. (Cited on pp. 134, 135)

[23] W. L. BRIGGS, V. E. HENSON, AND S. F. MCCORMICK, *A Multigrid Tutorial*, 2nd ed., SIAM, Philadelphia, 2000. (Cited on p. 100)

[24] P. N. BROWN AND A. C. HINDMARSH, *Matrix-free methods for stiff systems of ODE's*, SIAM J. Numer. Anal., 23 (1986), pp. 610–638. (Cited on p. 99)

[25] P. N. BROWN, A. C. HINDMARSH, AND L. R. PETZOLD, *Using Krylov methods in the solution of large-scale differential-algebraic systems*, SIAM J. Sci. Comp., 15 (1994), pp. 1467–1488. (Cited on p. 99)

[26] P. N. BROWN AND Y. SAAD, *Hybrid Krylov methods for nonlinear systems of equations*, SIAM J. Sci. Stat. Comp., 11 (1990), pp. 450–481. (Cited on p. 21)

[27] I. W. BUSBRIDGE, *The Mathematics of Radiative Transfer*, no. 50 in Cambridge Tracts, Cambridge University Press, Cambridge, 1960. (Cited on pp. 59, 152)

[28] S. L. CAMPBELL, I. C. F. IPSEN, C. T. KELLEY, AND C. D. MEYER, *GMRES and the minimal polynomial*, BIT, 36 (1996), pp. 664–675. (Cited on pp. 97, 98)

[29] E. CANCÈS AND C. L. BRIS, *Can we outperform the DIIS approach for electronic structure calculations?*, Int. J. Quantum Chem., 79 (2000), pp. 82–90. (Cited on p. 134)

[30] É. CANCÈS, G. KEMLIN, AND A. LEVITT, *Convergence analysis of direct minimization and self-consistent iterations*, SIAM J. Sci. Comp., 42 (2021), pp. 243–274. (Cited on p. 134)

[31] N. N. CARLSON AND K. MILLER, *Design and application of a gradient weighted moving finite element code I: In one dimension*, SIAM J. Sci. Comp., 19 (1998), pp. 728–765. (Cited on p. 133)

[32] S. CHANDRASEKHAR, *Radiative Transfer*, Dover, New York, 1960. (Cited on pp. 59, 92)

[33] X. CHEN AND C. T. KELLEY, *Convergence of the EDIIS algorithm for nonlinear equations*, SIAM J. Sci. Comp., 41 (2019), pp. A365–A379, https://doi.org/10.1137/18M1171084. (Cited on pp. 133, 136, 137, 139, 148)

[34] T. S. COFFEY, C. T. KELLEY, AND D. E. KEYES, *Pseudotransient continuation and differential-algebraic equations*, SIAM J. Sci. Comp., 25 (2003), pp. 553–569. (Cited on pp. 15, 16, 17, 18)

[35] T. S. COFFEY, R. J. MCMULLAN, C. T. KELLEY, AND D. S. MCRAE, *Globally convergent algorithms for nonsmooth nonlinear equations in computational fluid dynamics*, J. Comp. Appl. Math., 152 (2003), pp. 69–81. (Cited on pp. 17, 18)

[36] T. F. COLEMAN AND Y. LI, *A reflective Newton method for minimizing a quadratic function subject to bounds on some of the variables*, SIAM J. Optim., 6 (1996), pp. 1040–1058. (Cited on p. 148)

[37] T. F. COLEMAN AND J. J. MORÉ, *Estimation of sparse Jacobian matrices and graph coloring problems*, SIAM J. Numer. Anal., 20 (1983), pp. 187–209. (Cited on pp. 23, 56, 93)

[38] A. M. COLLIER, A. C. HINDMARSH, R. SERBAN, AND C. S. WOODWARD, *User documentation for KINSOL v2.8.0*, Tech. Report UCRL-SM-208116, Lawrence Livermore National Laboratory, 2015. (Cited on pp. 21, 135, 138)

[39] M. COVALT, *ILUZero.jl*, Julia Package, 2021, https://github.com/mcovalt/ILUZero.jl. (Cited on p. 100)

[40] M. G. CRANDALL AND P. H. RABINOWITZ, *Bifurcation from simple eigenvalues*, J. Funct. Anal., 8 (1971), pp. 321–340. (Cited on p. 158)

[41] A. R. CURTIS, M. J. D. POWELL, AND J. K. REID, *On the estimation of sparse Jacobian matrices*, J. Inst. Math. Appl., 13 (1974), pp. 117–119. (Cited on pp. 23, 56, 93)

[42] E. DARVE AND M. WOOTTERS, *Numerical Linear Algebra with Julia*, SIAM, Philadelphia, 2021. (Cited on p. xviii)

[43] P. DAVIS AND P. RABINOWITZ, *Abscissas and weights for Gaussian quadratures of high order*, J. Res. Natl. Bur. Stand., 56 (1956), pp. 35–37. (Cited on p. 154)

[44] T. A. DAVIS, *Algorithm 832: UMFPACK, an unsymmetric-pattern multifrontal method*, ACM Trans. Math. Softw., 30 (2004), pp. 196–199. (Cited on pp. 23, 68)

[45] T. A. DAVIS, *Direct Methods for Sparse Linear Systems*, no. 2 in Fundamentals of Algorithms, SIAM, Philadelphia, 2006. (Cited on pp. 23, 56, 68)

[46] H. DE STERCK AND Y. HE, *Anderson acceleration as a Krylov method with application to asymptotic convergence analysis*, arXiv:2109.14181v1 [math.NA], 2021, https://arxiv.org/abs/2109.14181. (Cited on p. 136)

[47] H. DE STERCK AND Y. HE, *Linear asymptotic convergence of Anderson acceleration: Fixed-point analysis*, arXiv:2109.14176v1 [math.OC], 2021, https://arxiv.org/abs/2109.14176. (Cited on p. 136)

[48] D. W. DECKER AND C. T. KELLEY, *Newton's method at singular points I*, SIAM J. Numer. Anal., 17 (1980), pp. 66–70. (Cited on pp. 25, 59, 92)

[49] D. W. DECKER AND C. T. KELLEY, *Sublinear convergence of the chord method at singular points*, Numer. Math., 42 (1983), pp. 147–154. (Cited on p. 92)

[50] P. H. DEDRICHS AND R. ZELLER, *Self-consistency iterations in electronic-structure calculations*, Phys. Rev. B, 28 (1983), pp. 5462–5472. (Cited on pp. 21, 133)

[51] J. DEGROOTE, K.-J. BATHE, AND J. VIERENDEELS, *Performance of a new partitioned procedure versus a monolithic procedure in fluid-structure interaction*, Comp. Structures, 97 (2009), pp. 793–801. (Cited on pp. 133, 134)

[52] R. S. DEMBO, S. C. EISENSTAT, AND T. STEIHAUG, *Inexact Newton methods*, SIAM J. Numer. Anal., 19 (1982), pp. 400–408. (Cited on p. 9)

[53] J. W. DEMMEL, *Applied Numerical Linear Algebra*, SIAM, Philadelphia, 1997. (Cited on pp. xii, 27, 50, 96, 97)

[54] J. E. DENNIS, JR., AND R. B. SCHNABEL, *Numerical Methods for Unconstrained Optimization and Nonlinear Equations*, no. 16 in Classics in Applied Mathematics, SIAM, Philadelphia, 1996. (Cited on pp. 2, 4, 12, 13, 15, 21, 25, 57)

[55] P. DEUFLHARD, *Adaptive pseudo-transient continuation for nonlinear steady state problems*, Tech. Report 02-14, Konrad-Zuse-Zentrum für Informationstechnik, Berlin, March 2002. (Cited on p. 15)

[56] E. J. DOEDEL, *Lecture Notes on Numerical Analysis of Bifurcation Problems*. Lecture notes from Sommerschule über Nichtlineare Gleichungssysteme, Hamburg, Germany, March 17–21, 1997. Available by anonymous ftp to ftp.cs.condordia.cainpub/doedel/doc/hamburg.ps.Z, 1997. (Cited on p. 157)

[57] J. J. DONGARRA, C. B. MOLER, J. R. BUNCH, AND G. W. STEWART, *LINPACK Users' Guide*, SIAM, Philadelphia, 1979. (Cited on pp. 27, 50, 51, 140)

[58] T. A. Driscoll and R. J. Braun, *Fundamentals of Numerical Computation: Julia Edition*, SIAM, Philadelphia, 2022. (Cited on p. xviii)

[59] S. C. EISENSTAT AND H. F. WALKER, *Globally convergent inexact Newton methods*, SIAM J. Optim., 4 (1994), pp. 393–422. (Cited on p. 13)

[60] S. C. EISENSTAT AND H. F. WALKER, *Choosing the forcing terms in an inexact Newton method*, SIAM J. Sci. Comp., 17 (1996), pp. 16–32. (Cited on pp. 10, 102, 103)

[61] P. ETTENHUBER AND P. JØRGENSEN, *Discarding information from previous iterations in an optimal way to solve the coupled cluster amplitude equations*, J. Chem. Theory Comp., 11 (2015), pp. 1518–1524, https://doi.org/10.1021/ct501114q. (Cited on p. 133)

[62] C. EVANS, S. POLLOCK, L. G. REBHOLZ, AND M. XIAO, *A proof that Anderson acceleration improves the convergence rate in linearly converging fixed-point methods (but not in those converging quadratically)*, SIAM J. Numer. Anal., 58 (2020), pp. 788–810. (Cited on p. 136)

[63] H.-R. FANG AND Y. SAAD, *Two classes of multisecant methods for nonlinear acceleration*, Numer. Linear Algebra Appl., 16 (2009), pp. 197–221, https://doi.org/10.1002/nla. (Cited on p. 136)

[64] M. W. FARTHING, C. E. KEES, T. COFFEY, C. T. KELLEY, AND C. T. MILLER, *Efficient steady-state solution techniques for variably saturated groundwater flow*, Adv. Water Resour., 26 (2003), pp. 833–849. (Cited on p. 18)

[65] K. R. FOWLER AND C. T. KELLEY, *Pseudo-transient continuation for nonsmooth nonlinear equations*, SIAM J. Numer. Anal., 43 (2005), pp. 1385–1406. (Cited on pp. 16, 17, 18)

[66] R. W. FREUND, *A transpose-free quasi-minimal residual algorithm for non-Hermitian linear systems*, SIAM J. Sci. Comp., 14 (1993), pp. 470–482. (Cited on p. 98)

[67] M. FRIGO AND S. G. JOHNSON, *The design and implementation of FFTW3*, Proc. IEEE, 93 (2005), pp. 216–231, https://doi.org/10.1109/JPROC.2004.840301. Special issue on "Program Generation, Optimization, and Platform Adaptation." (Cited on p. 59)

[68] M. J. FRISCH, G. W. TRUCKS, H. B. SCHLEGEL, G. E. SCUSERIA, M. A. ROBB, J. R. CHEESEMAN, G. SCALMANI, V. BARONE, G. A. PETERSSON, H. NAKATSUJI, X. LI, M. CARICATO, A. V. MARENICH, J. BLOINO, B. G. JANESKO, R. GOMPERTS, B. MENNUCCI, H. P. HRATCHIAN, J. V. ORTIZ, A. F. IZMAYLOV, J. L. SONNENBERG, D. WILLIAMS-YOUNG, F. DING, F. LIPPARINI, F. EGIDI, J. GOINGS, B. PENG, A. PETRONE, T. HENDERSON, D. RANASINGHE, V. G. ZAKRZEWSKI, J. GAO, N. REGA, G. ZHENG, W. LIANG, M. HADA, M. EHARA, K. TOYOTA, R. FUKUDA, J. HASEGAWA, M. ISHIDA, T. NAKAJIMA, Y. HONDA, O. KITAO, H. NAKAI, T. VREVEN, K. THROSSELL, J. A. MONTGOMERY, JR., J. E. PERALTA, F. OGLIARO, M. J. BEARPARK, J. J. HEYD, E. N. BROTHERS, K. N. KUDIN, V. N. STAROVEROV, T. A. KEITH, R. KOBAYASHI, J. NORMAND, K. RAGHAVACHARI, A. P. RENDELL, J. C. BURANT, S. S. IYENGAR, J. TOMASI, M. COSSI, J. M. MILLAM, M. KLENE, C. ADAMO, R. CAMMI, J. W. OCHTERSKI, R. L. MARTIN, K. MOROKUMA, O. FARKAS, J. B. FORESMAN, AND D. J. FOX, *Gaussian 16 Revision A.03*, Gaussian Inc., Wallingford, CT, 2016. (Cited on pp. 134, 137)

[69] V. GANINE, U. JAVIYA, N. HILLS, AND J. CHEW, *Coupled fluid-structure transient thermal analysis of a gas turbine internal air system with multiple cavities*, J. Eng. Gas Turbines Power, 134 (2012), p. 102508, https://doi.org/10.1115/1.4007060. (Cited on p. 134)

[70] D. J. GARDNER, C. S. WOODWARD, D. R. REYNOLDS, G. HOMMES, S. AUBREY, AND A. ARSNELIS, *Implicit integration methods for dislocation dynamics*, Modelling Simul. Mater. Sci. Eng., 23 (2015), 025006. (Cited on p. 134)

[71] A. J. GARZA AND G. E. SCUSERIA, *Comparison of self-consistent field convergence acceleration techniques*, J. Chem. Phys., 137 (2012), 054110, https://doi.org/10.1063/1.4740249. (Cited on p. 133)

[72] C. W. GEAR, *Numerical Initial Value Problems in Ordinary Differential Equations*, Prentice-Hall, Englewood Cliffs, NJ, 1971. (Cited on p. 17)

[73] M. GEE, C. T. KELLEY, AND R. B. LEHOUCQ, *Pseudo-transient continuation for nonlinear transient elasticity*, Int. J. Numer. Methods Eng., 78 (2009), pp. 1209–1219. (Cited on p. 18)

[74] G. H. GOLUB AND M. A. SAUNDERS, *Linear least squares and quadratic programming*, Tech. Report CS 134, Stanford University, 1969. (Cited on p. 148)

[75] G. H. GOLUB AND C. F. VAN LOAN, *Matrix Computations*, Johns Hopkins Studies in the Mathematical Sciences, 3rd ed., Johns Hopkins University Press, Baltimore, 1996. (Cited on pp. 50, 59)

[76] W. J. F. GOVAERTS, *Numerical Methods for Bifurcations of Dynamical Equilibria*, SIAM, Philadelphia, 2000. (Cited on pp. 46, 157)

[77] M. GRANT AND S. BOYD, *CVX: Matlab software for disciplined convex programming, version 2.0.* http://cvxr.com/cvx, Aug. 2012. (Cited on p. 135)

[78] K.-T. GRASSER, *Mixed-mode device simulation*, Ph.D. thesis, Technical University of Vienna, 1999, http://www.iue.tuwien.ac.at/publications/PhDTheses/grasser/diss.html. (Cited on p. 18)

[79] A. GREENBAUM, *Iterative Methods for Solving Linear Systems*, no. 17 in Frontiers in Applied Mathematics, SIAM, Philadelphia, 1997. (Cited on pp. 98, 99)

[80] A. GRIEWANK, *Evaluating Derivatives: Principles and Techniques of Algorithmic Differentiation*, no. 19 in Frontiers in Applied Mathematics, SIAM, Philadelphia, 2000. (Cited on pp. 5, 55, 92)

[81] R. HAELTERMAN, J. DEGROOTE, D. VAN HEULE, AND J. VIERENDEELS, *The quasi-Newton least squares method: A new and fast secant method analyzed for linear systems*, SIAM J. Numer. Anal., 47 (2009), pp. 2347–2368. (Cited on pp. 133, 134)

[82] S. HAMILTON, M. BERRILL, K. CLARNO, R. PAWLOWSKI, A. TOTH, C. T. KELLEY, T. EVANS, AND B. PHILIP, *An assessment of coupling algorithms for nuclear reactor core physics simulations*, J. Comp. Phys., 311 (2016), pp. 241–257. (Cited on pp. 24, 134)

[83] M. HATHERLY, M. PIIBELEHT, F. EKRE, AND OTHER CONTRIBUTORS, *Documenter.jl*, 2021, https://github.com/JuliaDocs/Documenter.jl. Julia Package. (Cited on p. xix)

[84] M. F. HERBST AND A. LEVITT, *Black-box inhomogeneous preconditioning for self-consistent field iterations in density functional theory*, J. Phys. Condens. Matter, 33 (2021), 085503, https://doi.org/10.1088/1361-648X/abcbdb. (Cited on pp. 133, 134)

[85] M. F. HERBST, A. LEVITT, AND E. CANCÈS, *DFTK: A Julian approach for simulating electrons in solids*, Proc. JuliaCon Conf., 3 (2021), 69, https://doi.org/10.21105/jcon.00069. (Cited on p. 134)

[86] M. A. HEROUX, R. A. BARTLETT, V. E. HOWLE, R. J. HOEKSTRA, J. J. HU, T. G. KOLDA, R. B. LEHOUCQ, K. R. LONG, R. P. PAWLOWSKI, E. T. PHIPPS, A. G. SALINGER, H. K. THORNQUIST, R. S. TUMINARO, J. M. WILLENBRING, A. WILLIAMS, AND K. S. STANLEY, *An overview of the Trilinos project*, ACM Trans. Math. Softw., 31 (2005), pp. 397–423. (Cited on pp. xii, 21, 55, 104)

[87] M. R. HESTENES AND E. STEIFEL, *Methods of conjugate gradient for solving linear systems*, J. Res. Natl. Bur. Stand., 49 (1952), pp. 409–436. (Cited on p. 98)

[88] D. J. HIGHAM, *Trust region algorithms and timestep selection*, SIAM J. Numer. Anal., 37 (1999), pp. 194–210. (Cited on p. 15)

[89] D. J. HIGHAM AND N. J. HIGHAM, *MATLAB Guide*, 2nd ed., SIAM, Philadelphia, 2005. (Cited on p. xvii)

[90] N. J. HIGHAM, *Accuracy and Stability of Numerical Algorithms*, SIAM, Philadelphia, PA, 1996. (Cited on p. xii)

[91] A. C. HINDMARSH, P. N. BROWN, K. E. GRANT, S. L. LEE, R. SERBAN, D. E. SHUMAKER, AND C. S. WOODWARD, *SUNDIALS: Suite of nonlinear and differential/algebraic equation solvers*, ACM Trans. Math. Softw., 31 (2005), pp. 363–396. (Cited on pp. xii, 21)

[92] T. HOLY, *Revise.jl*, 2020, https://github.com/timholy/Revise.jl. Julia Package. (Cited on pp. xiv, 40)

[93] S. HØST, J. OLSEN, B. JANSÍK, L. THØGERSEN, P. JØRGENSEN, AND T. HELGAKER, *The augmented Roothaan-Hall method for optimizing Hartree-Fock and Kohn-Sham density matrices*, J. Chem. Phys., 129 (2008), 124106, https://doi.org/10.1063/1.2974099. (Cited on p. 133)

[94] P. HOVLAND AND B. NORRIS, *Argonne National Laboratory Computational Differentiation Project*, 2002. http://www.mcs.anl.gov/autodiff/. (Cited on pp. 5, 55)

[95] W. HU, L. LIN, AND C. YANG, *Projected commutator DIIS method for accelerating hybrid functional electronic structure calculations*, J. Chem. Theory Comp., 13 (2017), pp. 5458–5467, https://doi.org/10.1021/acs.jctc.7b00892. (Cited on p. 133)

[96] IEEE COMPUTER SOCIETY, *IEEE standard for binary floating point arithmetic, IEEE Std 754-1885*, 1985. (Cited on p. 5)

[97] IEEE COMPUTER SOCIETY, *IEEE standard for floating-point arithmetic, IEEE Std 754-2019*, July 2019. (Cited on p. 5)

[98] M. INNES, *Don't unroll adjoint: Differentiating SSA-form programs*, arXiv:1810.07951, 2018, http://arxiv.org/abs/1810.07951. (Cited on pp. 5, 55, 92)

[99] I. C. F. IPSEN, *Numerical Matrix Analysis*, SIAM, Philadelphia, 2009. (Cited on p. xii)

[100] S. G. JOHNSON, *QuadGK.jl: Gauss–Kronrod integration in Julia*, Julia Package, 2013, https://github.com/JuliaMath/QuadGK.jl. (Cited on p. 154)

[101] S. G. JOHNSON, *LaTeXStrings.jl*, Julia Package, 2017, https://github.com/stevengj/LaTeXStrings.jl. (Cited on p. 3)

[102] S. G. JOHNSON, *FFTW.jl*, Julia Package, 2020, https://github.com/JuliaMath/FFTW.jl. (Cited on pp. 59, 104)

[103] T. KANT AND S. PATEL, *Transient/pseudo-transient finite element small/large deformation analysis of two-dimensional problems*, Comp. Structures, 36 (1990), pp. 421–427. (Cited on p. 18)

[104] S. KARLIN, *Positive operators*, J. Math. Mech., 8 (1959), pp. 907–937. (Cited on p. 159)

[105] H. B. KELLER, *Newton's method under mild differentiability conditions*, J. Comp. Sys. Sci, 4 (1970), pp. 15–28. (Cited on pp. 24, 48)

[106] H. B. KELLER, *Lectures on Numerical Methods in Bifurcation Theory*, Tata Institute of Fundamental Research, Lectures on Mathematics and Physics, Springer-Verlag, New York, 1987. (Cited on pp. 46, 59, 65, 157, 159, 161, 162)

[107] C. T. KELLEY, *Iterative Methods for Linear and Nonlinear Equations*, no. 16 in Frontiers in Applied Mathematics, SIAM, Philadelphia, 1995. (Cited on pp. xi, xii, 2, 4, 8, 9, 10, 12, 13, 15, 21, 22, 26, 53, 60, 67, 96, 97, 98, 99, 102, 138)

[108] C. T. KELLEY, *Multilevel source iteration accelerators for the linear transport equation in slab geometry*, Transp. Theory Stat. Phys., 24 (1995), pp. 679–708. (Cited on pp. 153, 154)

[109] C. T. KELLEY, *Existence and uniqueness of solutions of nonlinear systems of conductive-radiative heat transfer equations*, Transp. Theory Stat. Phys., 25 (1996), pp. 249–260. (Cited on pp. 149, 150)

[110] C. T. KELLEY, *Iterative Methods for Optimization*, no. 18 in Frontiers in Applied Mathematics, SIAM, Philadelphia, 1999. (Cited on p. 2)

[111] C. T. KELLEY, *Solving Nonlinear Equations with Newton's Method*, no. 1 in Fundamentals of Algorithms, SIAM, Philadelphia, 2003. (Cited on pp. xi, xii, 14, 21, 56, 59, 60, 67, 93, 101)

[112] C. T. KELLEY, *Numerical methods for nonlinear equations*, Acta Numerica, 27 (2018), pp. 207–287, https://doi.org/10.1017/S0962492917000113. (Cited on pp. 2, 16, 59)

[113] C. T. KELLEY, *Newton's method in mixed precision*, SIAM Review, 64 (2022), pp. 191–211, https://doi.org/10.1137/20M1342902. (Cited on pp. 6, 50, 59, 79, 92)

[114] C. T. KELLEY, *Notebook for Solving Nonlinear Equations with Iterative Methods: Solvers and Examples in Julia*, IJulia Notebook, 2022, https://doi.org/10.5281/zenodo.4284687, https://github.com/ctkelley/NotebookSIAMFANL. (Cited on p. xi)

[115] C. T. KELLEY, *SIAMFANLEquations.jl*, Julia Package, 2022, https://doi.org/10.5281/zenodo.4284687, https://github.com/ctkelley/SIAMFANLEquations.jl. (Cited on p. xi)

[116] C. T. KELLEY AND D. E. KEYES, *Convergence analysis of pseudo-transient continuation*, SIAM J. Numer. Anal., 35 (1998), pp. 508–523. (Cited on pp. 15, 16, 17, 19, 26)

[117] C. T. KELLEY, L.-Z. LIAO, L. QI, M. T. CHU, J. P. REESE, AND C. WINTON, *Projected pseudotransient continuation*, SIAM J. Numer. Anal., 46 (2008), pp. 3071–3083. (Cited on p. 18)

[118] C. T. KELLEY, C. T. MILLER, AND M. D. TOCCI, *Termination of Newton/chord iterations and the method of lines*, SIAM J. Sci. Comp., 19 (1998), pp. 280–290. (Cited on pp. 11, 66)

[119] C. T. KELLEY AND E. W. SACHS, *Mesh independence of Newton-like methods for infinite dimensional problems*, J. Integr. Equations Appl., 3 (1991), pp. 549–573. (Cited on p. 59)

[120] T. KERKHOVEN AND Y. SAAD, *On acceleration methods for coupled nonlinear elliptic systems*, Numer. Math., 60 (1992), pp. 525–548. (Cited on pp. 6, 16)

[121] D. E. KEYES, *Aerodynamic applications of Newton-Krylov-Schwarz solvers*, in Proceedings of the 14th International Conference on Numerical Methods in Fluid Dynamics, R. Narasimha, eds., Springer, New York, 1995, pp. 1–20. (Cited on p. 18)

[122] D. E. KEYES AND M. D. SMOOKE, *A parallelized elliptic solver for reacting flows*, in Parallel Computations and Their Impact on Mechanics, A. K. Noor, ed., ASME, New York, 1987, pp. 375–402. (Cited on pp. 16, 18)

[123] D. A. KNOLL AND W. J. RIDER, *A multigrid preconditioned Newton-Krylov method*, Tech. Report LA-UR-97-4013, Los Alamos National Laboratory, 1997. (Cited on p. 18)

[124] K. N. KUDIN, G. E. SCUSERIA, AND E. CANCÈS, *A black-box self-consistent field convergence algorithm: One step closer*, J. Chem. Phys., 116 (2002), pp. 8255–8261, https://doi.org/10.1063/1.1470195. (Cited on pp. 133, 134, 136, 137, 148)

[125] Y. A. KUZNETSOV, *Elements of Applied Bifurcation Theory*, Springer, New York, 1998. (Cited on p. 157)

[126] E. W. LARSEN AND P. NELSON, *Finite difference approximations and superconvergence for the discrete ordinate equations in slab geometry*, SIAM J. Numer. Anal., 19 (1982), pp. 334–348. (Cited on p. 153)

[127] B. LAUENS AND A. DOWNEY, *Think Julia: How to Think Like a Computer Scientist*, O'Reilly Media, Sebastpol, CA, 2019, https://benlauwens.github.io/ThinkJulia.jl/latest/book.html. (Cited on p. xviii)

[128] L. LIN AND J. LU, *A Mathematical Introduction to Electronic Structure Theory*, no. 4 in SIAM Spotlights, SIAM, Philadelphia, 2019. (Cited on p. 21)

[129] L. LIN AND C. YANG, *Elliptic preconditioner for accelerating the self-consistent field iteration in Kohn–Sham density functional theory*, SIAM J. Sci. Comp., 35 (2013), pp. S277–S298. (Cited on p. 133)

[130] F. LINDNER, M. MEHL, K. SCHEUFELE, AND B. UEKERMANN, *A comparison of various quasi-newton schemes for partitioned fluid-structure interaction*, in ECCOMAS Coupled Problems in Science and Engineering, Venice, B. A. Schrefler, E. Oñate, and M. Papadrakakis, eds., DIMNE, Barcelona, 2015, pp. 477–488. (Cited on pp. 133, 134)

[131] P. A. LOTT, H. F. WALKER, C. S. WOODWARD, AND U. M. YANG, *An accelerated Picard method for nonlinear systems related to variably saturated flow*, Adv. Water Resour., 38 (2012), pp. 92–101. (Cited on p. 134)

[132] J. N. LYNESS AND C. B. MOLER, *Numerical differentiation of analytic functions*, SIAM J. Numer. Anal., 4 (1967), pp. 202–210. (Cited on p. 55)

[133] M. N. ÖZIŞIK, *Radiative Transfer and Interaction with Conduction and Convection*, John Wiley and Sons, New York, 1973. (Cited on pp. 149, 150)

[134] T. A. MANTEUFFEL AND S. V. PARTER, *Preconditioning and boundary conditions*, SIAM J. Numer. Anal., 27 (1990), pp. 656–694. (Cited on p. 100)

[135] J. MARSDEN AND T. J. R. HUGHES, *Mathematical Foundations of Elasticity*, Dover, New York, 1983. (Cited on p. 65)

[136] K. MILLER, *Nonlinear Krylov and moving nodes in the method of lines*, J. Comp. Appl. Math., 183 (2005), pp. 275–287. (Cited on p. 133)

[137] P. MOGENSEN, *Nlsolve.jl*, Julia Package, 2020, https://github.com/JuliaNLSolvers/NLsolve.jl. (Cited on p. 21)

[138] S. MOORE, E. BRIGGS, M. HODAK, W. LU, J. BERNHOLC, AND C. LEE, *Scaling the RMG quantum mechanics code*, in Proceedings of the Extreme Scaling Workshop, BW-XSEDE '12, Champaign, IL, 2012, University of Illinois at Urbana-Champaign, pp. 8:1–8:6. (Cited on pp. 134, 135)

[139] J. J. MORÉ, B. S. GARBOW, AND K. E. HILLSTROM, *User guide for MINPACK-1*, Tech. Report ANL-80-74, Argonne National Laboratory, 1980. (Cited on p. 21)

[140] W. MULDER AND B. V. LEER, *Experiments with implicit upwind methods for the Euler equations*, J. Comp. Phys., 59 (1985), pp. 232–246. (Cited on p. 18)

[141] T. W. MULLIKIN, *Some probability distributions for neutron transport in a half space*, J. Appl. Prob., 5 (1968), pp. 357–374. (Cited on p. 59)

[142] N. M. NACHTIGAL, S. C. REDDY, AND L. N. TREFETHEN, *How fast are nonsymmetric matrix iterations?*, SIAM J. Matrix Anal. Appl., 13 (1992), pp. 778–795. (Cited on p. 97)

[143] R. M. NIXON, *Conversation with John Dean*, 1972. (Cited on pp. 55, 129)

[144] S. OLVER, *BandedMatrices.jl*, Julia Package, 2020, `https://github.com/JuliaMatrices/BandedMatrices.jl`. (Cited on pp. 23, 62, 63, 83)

[145] S. OLVER, *FastGaussQuadrature.jl*, Julia Package, 2021, `https://github.com/JuliaApproximation/FastGaussQuadrature.jl`. (Cited on p. 154)

[146] C. W. OOSTERLEE AND T. WASHIO, *Krylov subspace acceleration for nonlinear multigrid with application to recirculating flows*, SIAM J. Sci. Comp., 21 (2000), pp. 1670–1690. (Cited on p. 133)

[147] P. D. ORKWIS AND D. S. MCRAE, *Newton's method solver for high-speed separated flow-fields*, AIAA J., 30 (1992), pp. 78–85. (Cited on p. 18)

[148] P. D. ORKWIS AND D. S. MCRAE, *Newton's method solver for the axisymmetric Navier-Stokes equations*, AIAA J., 30 (1992), pp. 1507–1514. (Cited on p. 18)

[149] J. M. ORTEGA AND W. C. RHEINBOLDT, *Iterative Solution of Nonlinear Equations in Several Variables*, Academic Press, New York, 1970. (Cited on pp. 4, 12, 13, 15, 133)

[150] M. L. OVERTON, *Numerical Computing with IEEE Floating Point Arithmetic*, SIAM, Philadelphia, 2001. (Cited on p. 5)

[151] M. PERNICE AND H. F. WALKER, *NITSOL: A Newton iterative solver for nonlinear systems*, SIAM J. Sci. Comp., 19 (1998), pp. 302–318. (Cited on p. 21)

[152] J. PITKÄRANTA AND L. R. SCOTT, *Error estimates for the combined spatial and angular approximations of the transport equation for slab geometry*, SIAM J. Numer. Anal., 20 (1983), pp. 922–950. (Cited on p. 153)

[153] S. POLLOCK AND L. G. REBHOLZ, *Anderson acceleration for contractive and noncontractive operators*, IMA J. Numer. Anal., 41 (2021), pp. 2841–2872, `https://doi.org/10.1093/imanum/draa095`. Published online Jan 6, 2021. (Cited on pp. 134, 136, 148)

[154] S. POLLOCK, L. G. REBHOLZ, AND M. XIAO, *Anderson-accelerated convergence of Picard iterations for incompressible Navier–Stokes equations*, SIAM J. Numer. Anal., 57 (2019), pp. 615–637, `https://doi.org/10.1137/18M1206151`. (Cited on pp. 134, 136)

[155] F. A. POTRA AND H. ENGLER, *A characterization of the behavior of the Anderson acceleration on linear problems*, Linear Algebra Appl., 438 (2012), pp. 1002–1011. (Cited on p. 136)

[156] M. J. D. POWELL, *A hybrid method for nonlinear equations*, in Numerical Methods for Nonlinear Algebraic Equations, P. Rabinowitz, ed., Gordon and Breach, New York, 1970, pp. 87–114. (Cited on pp. 15, 21, 25)

[157] P. PULAY, *Improved SCF convergence acceleration*, J. Comp. Chem., 3 (1982), pp. 556–560. (Cited on p. 133)

[158] L. QI AND J. SUN, *A nonsmooth version of Newton's method*, Math. Program., 58 (1993), pp. 353–367. (Cited on p. 24)

[159] A. QUARTERONI AND A. VALLI, *Domain Decomposition Methods for Partial Differential Equations*, Oxford University Press, London, 1999. (Cited on p. 100)

[160] C. RACKAUCKAS, P. MISHRA, S. GOWDA, AND L. HUANG, *SparseDiffTools.jl*, Julia Package, 2020, `https://github.com/JuliaDiff/SparseDiffTools.jl`. (Cited on pp. 23, 56, 93)

[161] C. RACKAUCKAS AND Q. NIE, *DifferentialEquations.jl–a performant and feature-rich ecosystem for solving differential equations in Julia*, J. Open Res. Softw., 5 (2017), 15. (Cited on p. 21)

[162] K. RADHAKRISHNAN AND A. C. HINDMARSH, *Description and use of LSODE, the Livermore solver for ordinary differential equations*, Tech. Report URCL-ID-113855, Lawrence Livermore National Laboratory, December 1993. (Cited on pp. 11, 53, 66)

[163] J. REVELS, *ReverseDiff.jl*, Julia Package, 2020, `https://github.com/JuliaDiff/ReverseDiff.jl`. Julia Package. (Cited on p. 55)

[164] J. REVELS, M. LUBIN, AND T. PAPAMARKOU, *Forward-mode automatic differentiation in Julia*, arXiv:1607.07892 [cs.MS], 2016, `https://arxiv.org/abs/1607.07892`. (Cited on pp. 5, 55, 92)

[165] W. C. RHEINBOLDT, *Numerical Analysis of Parametrized Nonlinear Equations*, John Wiley and Sons, New York, 1986. (Cited on pp. 157, 161)

[166] T. ROHWEDDER AND R. SCHNEIDER, *An analysis for the DIIS acceleration method used in quantum chemistry calculations*, J. Math. Chem., 49 (2011), pp. 1889–1914, `http://www.springerlink.com/index/10.1007/s10910-011-9863-y`. (Cited on pp. 133, 136)

[167] Y. SAAD, *ILUM: A multi-elimination ILU preconditioner for general sparse matrices*, SIAM J. Sci. Comp., 17 (1996), pp. 830–847. (Cited on pp. 24, 100)

[168] Y. SAAD, *Iterative Methods for Sparse Linear Systems*, Prindle, Weber, and Schmidt, New York, 1996. (Cited on p. 100)

[169] Y. SAAD, J. R. CHELIKOWSKY, AND S. M. SHONTZ, *Numerical methods for electronic structure calculations of materials*, SIAM Review, 52 (2010), pp. 3–54. (Cited on pp. 21, 136)

[170] Y. SAAD AND M. H. SCHULTZ, *GMRES: A generalized minimal residual algorithm for solving nonsymmetric linear systems*, SIAM J. Sci. Stat. Comp., 7 (1986), pp. 856–869. (Cited on pp. 10, 96, 138)

[171] A. G. SALINGER, N. M. BOU-RABEE, R. P. PAWLOWSKI, E. D. WILKES, E. A. BURROUGHS, R. B. LEHOUCQ, AND L. A. ROMERO, *LOCA 1.0 Library of Continuation Algorithms: Theory and Implementation Manual*, Tech. Report SAND2002-0396, Sandia National Laboratory, March 2002. (Cited on pp. 157, 162)

[172] H. B. SCHLEGEL AND J. J. W. MCDOUALL, *Do you have SCF stability and convergence problems?*, in Computational Advances in Organic Chemistry: Molecular Structure and Reactivity, C. Ögretir and I. G. Csizmadia, eds., Kluwer, Dordrecht, 1991, pp. 167–185. (Cited on p. 137)

[173] R. B. SCHNABEL, J. E. KOONTZ, AND B. E. WEISS, *A modular system of algorithms for unconstrained minimization*, ACM Trans. Math. Softw., 11 (1985), pp. 419–440, `ftp://ftp.cs.colorado.edu/users/uncmin/tape.jan30/shar`. (Cited on p. 21)

[174] R. SCHNEIDER, T. ROHWEDDER, A. NEELOV, AND J. BLAUERT, *Direct minimization for calculating invariant subspaces in density functional computations of the electronic structure*, J. Comp. Math., 27 (2008), pp. 360–387. (Cited on p. 133)

[175] V. E. SHAMANSKII, *A modification of Newton's method*, Ukr. Mat. Zh., 19 (1967), pp. 133–138 (in Russian). (Cited on p. 54)

[176] L. F. SHAMPINE, *Implementation of implicit formulas for the solution of ODEs*, SIAM J. Sci. Stat. Comp., 1 (1980), pp. 103–118. (Cited on p. 66)

[177] L. F. SHAMPINE, *Numerical Solution of Ordinary Differential Equations*, Chapman and Hall, New York, 1994. (Cited on p. 66)

[178] A. I. SHESTAKOV AND J. L. MILOVICH, *Applications of pseudo-transient continuation and Newton-Krylov methods for the Poisson-Boltzmann and radiation diffusion equations*, Tech. Report UCRL-JC-139339, Lawrence Livermore National Laboratory, 2000. (Cited on p. 18)

[179] R. SIEGEL AND J. R. HOWELL, *Thermal Radiation Heat Transfer*, 3rd ed., Hemisphere Publishing, Washington, 1992. (Cited on pp. 149, 150)

[180] C. E. SIEWERT, *An improved iterative method for solving a class of coupled conductive-radiative heat transfer problems*, J. Quant. Spectrosc. Radiat. Transfer, 54 (1995), pp. 599–605. (Cited on p. 149)

[181] C. E. SIEWERT AND J. R. THOMAS, *A computational method for solving a class of coupled conductive-radiative heat transfer problems*, J. Quant. Spectrosc. Radiat. Transfer, 45 (1991), pp. 273–281. (Cited on pp. 149, 155, 163)

[182] B. SMITH, P. BJØRSTAD, AND W. GROPP, *Domain Decomposition: Parallel Multilevel Methods for Elliptic Partial Differential Equations*, Cambridge University Press, Cambridge, 1996. (Cited on p. 100)

[183] M. D. SMOOKE, R. MITCHELL, AND D. KEYES, *Numerical solution of two-dimensional axisymmetric laminar diffusion flames*, Combust. Sci. Technol., 67 (1989), pp. 85–122. (Cited on p. 18)

[184] W. SQUIRE AND G. TRAPP, *Using complex variables to estimate derivatives of real functions*, SIAM Review, 40 (1998), pp. 110–112. (Cited on p. 55)

[185] G. W. STEWART, *Introduction to Matrix Computations*, Academic Press, New York, 1973. (Cited on p. 50)

[186] H. STOPPELS, *IncompleteLU.jl*, Julia Package, 2021, `https://github.com/haampie/IncompleteLU.jl`. (Cited on p. 100)

[187] A. G. TAYLOR AND A. C. HINDMARSH, *User documentation for KINSOL, a nonlinear solver for sequential and parallel computers*, Tech. Report UCRL-ID-131185, Lawrence Livermore National Laboratory, Center for Applied Scientific Computing, July 1998. (Cited on p. 21)

[188] THE JULIA PROJECT, *The Julia Language*, 2020, `https://docs.julialang.org/en/v1/`. (Cited on pp. xviii, 14)

[189] A. TOSELLI AND O. WIDLUND, *Domain Decomposition Methods — Algorithms and Theory*, Springer, Berlin, 2005. (Cited on p. 100)

[190] A. TOTH, *A Theoretical Analysis of Anderson Acceleration and Its Application in Multiphysics Simulation for Light-Water Reactors*, Ph.D. thesis, North Carolina State University, Raleigh, North Carolina, 2016. (Cited on pp. 134, 135, 138)

[191] A. TOTH, J. A. ELLIS, T. EVANS, S. HAMILTON, C. T. KELLEY, R. PAWLOWSKI, AND S. SLATTERY, *Local improvement results for Anderson acceleration with inaccurate function evaluations*, SIAM J. Sci. Comp., 39 (2017), pp. S47–S65, https://doi.org/10.1137/16M1080677. (Cited on pp. 134, 136)

[192] A. TOTH AND C. T. KELLEY, *Convergence analysis for Anderson acceleration*, SIAM J. Numer. Anal., 53 (2015), pp. 805–819, https://doi.org/10.1137/130919398. (Cited on pp. 134, 135, 136, 141)

[193] A. TOTH, C. T. KELLEY, S. SLATTERY, S. HAMILTON, K. CLARNO, AND R. PAWLOWSKI, *Analysis of Anderson acceleration on a simplified neutronics/thermal hydraulics system*, in 2015 Joint International Conference on Mathematics and Computation (M&C), Supercomputing in Nuclear Applications (SNA) and the Monte Carlo (MC) Method, American Nuclear Society, Lagrange Park, IL, 2015. (Cited on p. 134)

[194] A. TOTH AND R. PAWLOWSKI, *NOX::Solver::AndersonAcceleration Class Reference*, 2015, https://trilinos.org/docs/dev/packages/nox/doc/html/classNOX_1_1Solver_1_1AndersonAcceleration.html. (Cited on pp. 135, 138)

[195] L. N. TREFETHEN AND D. BAU, *Numerical Linear Algebra*, SIAM, Philadelphia, 1996. (Cited on pp. xii, 50, 96, 97)

[196] H. UECKER, *Numerical Continuation and Bifurcation in Nonlinear PDEs*, SIAM, Philadelphia, 2021. (Cited on p. 157)

[197] H. A. VAN DER VORST, *Bi-CGSTAB: A fast and smoothly converging variant to Bi-CG for the solution of nonsymmetric systems*, SIAM J. Sci. Stat. Comp., 13 (1992), pp. 631–644. (Cited on pp. 96, 98)

[198] R. VELTZ, *BifurcationKit.jl*, Julia Package, 2020, https://github.com/rveltz/BifurcationKit.jl. (Cited on pp. 21, 157, 163)

[199] V. VENKATAKRISHNAN, *Newton solution of inviscid and viscous problems*, AIAA J., 27 (1989), pp. 885–891. (Cited on p. 18)

[200] R. VISKANTA, *Heat transfer by conduction and radiation in absorbing and scattering materials*, J. Heat Transfer, 87 (1965), pp. 143–150. (Cited on pp. 149, 150)

[201] H. W. WALKER AND P. NI, *Anderson acceleration for fixed-point iterations*, SIAM J. Numer. Anal., 49 (2011), pp. 1715–1735. (Cited on pp. 133, 135, 136, 138)

[202] T. WASHIO AND C. OOSTERLEE, *Krylov subspace acceleration for nonlinear multigrid schemes*, Elec. Trans. Numer. Anal., 6 (1997), pp. 271–290. (Cited on p. 133)

[203] L. T. WATSON, S. C. BILLUPS, AND A. P. MORGAN, *Algorithm 652: HOMPACK: A suite of codes for globally convergent homotopy algorithms*, ACM Trans. Math. Softw., 13 (1987), pp. 281–310. (Cited on pp. 15, 25)

[204] WIKI CONTRIBUTORS, *Julia for MATLAB Users*, 2020, https://en.wikibooks.org/wiki/Julia_for_MATLAB_Users. (Cited on p. xviii)

[205] J. WILLERT, W. T. TAITANO, AND D. KNOLL, *Leveraging Anderson acceleration for improved convergence of iterative solutions to transport systems*, J. Comp. Phys., 273 (2014), pp. 278–286, https://doi.org/10.1016/j.jcp.2014.05.015. (Cited on p. 134)

[206] U. M. YANG, *Parallel algebraic multigrid methods—high performance preconditioners*, in Numerical Solution of Partial Differential Equations on Parallel Computers, A. M. Bruaset and A. Tveito, eds., Springer, Berlin, 2006, pp. 209–236, `https://doi.org/10.1007/3-540-31619-1_6`. (Cited on p. 100)

[207] D. YOUNG, *Computational Chemistry: A Practical Guide for Applying Techniques to Real World Problems*, Wiley, New York, 2001. (Cited on p. 137)

[208] D. YOUNG, *SCF convergence and chaos theory*, 2001, `http://www.ccl.net/cca/documents/dyoung/topics-orig/converge.html`. (Cited on p. 137)

[209] J. ZHANG, B. O'DONOGHUE, AND S. BOYD, *Globally convergent Type-I Anderson acceleration for nonsmooth fixed-point iterations*, SIAM J. Optim., 30 (2020), pp. 3170–3197, `https://doi.org/10.1137/18M1232772`. (Cited on p. 136)

[210] M. ZIÓŁKOWSKI, V. WEIJO, P. JØRGENSEN, AND J. OLSEN, *An efficient algorithm for solving nonlinear equations with a minimal number of trial vectors: Applications to atomic-orbital based coupled-cluster theory*, J. Chem. Phys., 128 (2008), pp. 1–12, `https://doi.org/10.1063/1.2928803`. (Cited on p. 133)

Index